Contested Water

Urban and Industrial Environments
Series editor: Robert Gottlieb, Henry R. Luce Professor of Urban and Environmental Policy, Occidental College

For a complete list of books published in this series, please see the back of the book.

Contested Water

The Struggle Against Water Privatization in the
United States and Canada

Joanna L. Robinson

The MIT Press
Cambridge, Massachusetts
London, England

MIT Press books may be purchased at special quantity discounts for business or sales pro-motional use. For information, please email special_sales@mitpress.mit.edu or write to Spe-cial Sales Department, The MIT Press, 55 Hayward Street, Cambridge, MA 02142.

This book was set in Sabon by the MIT Press. Printed on recycled paper and bound in the United States of America.

Library of Congress Cataloging-in-Publication Data

Robinson, Joanna L., 1971-
Contested water : the struggle against water privatization in the United States and Canada / Joanna L. Robinson.
    pages   cm. — (Urban and industrial environments)
Includes bibliographical references and index.
ISBN 978-0-262-01885-2 (hardcover : alk. paper) — ISBN 978-0-262-51839-0 (pbk. :alk. paper)
1. Water utilities—United States. 2. Water utilities—Canada. 3. Municipal water supply—United States. 4. Municipal water supply—Canada. 5. Privatization—United States. 6. Privatization—Canada. I. Title.
HD4461.R63   2013
363.6'10973—dc23
2012031527

10   9   8   7   6   5   4   3   2   1

Contents

Preface

On a warm summer evening in June 2001, I attended a public meeting in the Canadian city of North Vancouver, British Columbia, along with hundreds of activists and members of the public. We were there to voice our opposition to a proposal by the Metro Vancouver Board to privatize the construction of the new Seymour water filtration plant, which would supply most of the drinking water for the region, a population base of more than two million. When I arrived at the meeting, there were literally hundreds of people pouring into the theater where the meeting was to be held. Many were dressed in costume, while others were holding banners and handing out pamphlets about the dangers of water privatization. Just outside the theater doors, a group of older women were singing songs chastising the local government for selling our water to the highest bidder. The atmosphere was festive, while at the same time I could sense a rising anger among the crowd as people filed in, greeted each other, and exchanged information. Soon most of the seats were full, and the meeting began.

The presenters—including senior Metro Vancouver bureaucrats and one elected official—began to speak, but were soon drowned out by a chorus of boos and shouts of "liars!" from the audience. With rising impatience, many people began demanding a chance for public input, stomping their feet and calling out "Let us speak!" When the time came for comments from the public, several dozen people, perhaps as many as a hundred, waited to speak, forming a line that snaked out the door. Almost all of the speakers were opposed to privatization. Their comments ranged from impassioned pleas to keep water part of the commons and angry derision about the lack of public input, to remarks on the legal risks under international trade agreements. After each public comment, the audience erupted in loud shouts of approval and intense applauding. As the two-hour meeting stretched to four and then five hours, with people still lined up to speak, the officials at the front of the room looked increasingly weary and uncomfortable.

Well past midnight, as the last comment was made and the meeting came to a close, with the Metro Vancouver representatives still seemingly entrenched in their position, the audience made its way out of the theater and into the warm summer night. Clearly, the momentum on the part of the audience would carry them to the next public meeting in Vancouver, scheduled for two weeks later. Although I was awed by the number of people who turned out for the meeting, and by their level of passion and commitment to keeping the water system public, I was surprised when, the next morning, while listening to the radio, I heard that the Metro Vancouver Board had reversed their decision to privatize the Seymour plant. They cited public opposition and the risks of contravening trade agreements.

How had anti–water privatization activists managed to mobilize so many people in such a short time and successfully reverse such an important policy decision? It was clear that the decision made by political elites had been reversed through the presentation of a powerful common message. I was curious to know more about how this happened and what made the movement so successful. That evening I began my journey to explore the politics of water and the growing conflict between the commons and commodification.

Why do people mobilize so passionately against the outsourcing of water services in their communities? What factors explain the trajectories of local anti–water privatization movements in an era of globalization? This book examines the struggles against water privatization that have emerged in communities around the world that challenge the marketization of water. The politics of water and the initiatives of social movements fighting to ensure protection of and fair access to water will be among the most important in human history. We all need water to survive. Many communities, particularly in developing countries, already experience severe water scarcity, while other areas are dealing with problems of pollution and failing infrastructure that are frequently seen in poor communities globally and—closer to home—occur all too often on First Nations reserves in Canada and on American Indian reservations in the United States.

In a world where the balance of power and material resources is more often than not tipped in favor of the wealthy and corporate interests and where a growing number of people lack access to life-sustaining water resources, it is no surprise that conflict over water is on the rise. A scarce resource, the ownership of water—who owns, manages, and has access to clean drinking water—will be critical for the lives of billions and will, ultimately, affect us all. Many activists argue that water is the next oil, and that the wars of the future will be fought over control and access to water. The battles have already begun, as communities around the world, from Cochabamba, Bolivia, to Atlanta, Georgia, and Stockton, California, have wrestled with the question of who should control their water. A matter of life and death, the politics of water, including the movements that mobilize to protect water as part of the commons, is one of the most critical, visible, and contested issues of our time.

Acknowledgments

This work would not have been possible without the help of many people to whom I owe a debt of gratitude. I especially want to thank the movement leaders who took time out of their busy lives to share their stories of activism. I was moved and inspired by their dedication to social and environmental justice, and heartened by their commitment to building better communities. This research would not have been possible without their generosity and willingness to meet with me for an interview.

I am grateful to the Social Sciences and Humanities Research Council of Canada (SSHRC) for funding the research for this book. A special note of thanks to David Tindall, whose encouragement and guidance from the initial idea to the final stages of the research helped improve the quality of the work. Several people also provided valuable feedback on early drafts of the work. I want to thank Jennifer Chun and Rima Wilkes for their guidance on the research design and data collection, and Jamie Peck and David Pellow for their suggestions on improving the work. I am especially grateful for the support and helpful advice of Clay Morgan and Robert Gottlieb at The MIT Press and to the anonymous reviewers for the Press, whose recommendations for revisions helped strengthen the book.

I also appreciate the encouragement of other scholars at the University of British Columbia who enthusiastically supported my research, including Ralph Matthews, Dawn Currie, Gerry Veenstra, Julian Dierkes, and Peter Dauvergne. As an undergraduate student at McGill University, I was fortunate to learn from and be inspired by many accomplished scholars. In particular, I want to thank Prue Rains and Karin Bauer, as well as Suzanne Staggenborg, who introduced me to scholarship on contentious politics and motivated me to complete a qualitative study of social movements. I am grateful, as well, for the support and mentorship of Kim Voss at the University of California, Berkeley. I also want to thank Jeffrey Cormier for his support of my work.

I am very fortunate to have a wonderful network of friends who have helped me keep balance in my life during the writing of this book. I thank Alan Jacobs and Antje Ellermann for their constant support and friendship, and for the many adventures we have shared over the years. I am grateful to Shannon Daub and Ryan Blogg for their friendship; our great conversations and laughter over delicious meals and lazy days on Bowen Island have been a delight in my life. For the past twenty-five years, Juliette Pelletier has been my closest friend and has supported me through thick and thin, helping me keep perspective and always giving me a reason to laugh. I thank her for her friendship. Trish Winston was my second mother during my undergraduate years in Montreal, and I am truly grateful for her kindness and generosity. I am very appreciative of Barbara and Syd Bulman-Fleming for their support and encouragement. I also thank Kyla Tienhaara and Kyle Horner for their friendship; their global adventures and commitment to environmental justice are inspiring. I also thank Sam Jones and Pierre Koch for their friendship, and for opening their home to us during our visits to Toronto. Veronique Sardi remains a wonderful friend and one of my greatest champions, and I have enjoyed her frequent visits to Vancouver. I greatly appreciate the support and friendship of my SPEC friends and colleagues Carole Christopher, Gerry Thorne, Dan Rogers, and Tara Moreau, whose passion for environmental justice is a constant source of inspiration. In addition, I thank Catherine Bischoff, Janelle Martin, Ken Hildebrand, Wendy Roth, Ian Tietjen, Amy Hanser, Nathan Lauster, Jeremy Weinstein, Rachel Gibson, and Molly and Ty Sterkel for their friendship and support.

My parents, Cam and Helen Robinson, always provided me steadfast support and encouragement of my academic pursuits as well as unconditional love, generosity, and friendship. I am grateful to my father for teaching me the importance of social justice from the beginning. I want to thank my mother, who recently passed away, to whom I owe an enormous debt of gratitude for showing me what matters most in life: love, friendship, honesty, perspective, and a sense of humor!

I thank my sister Leslie for her support and for always believing in me even when I doubted myself. My sister Michelle has been one of my biggest champions throughout my life, and I am grateful for her love and friendship. I also want to thank my brother-in-law Charles for the many delicious meals over the years and for being such a wonderful host during our many trips to Gatineau. My nieces, Kate and Elise, and nephews, Georges and Graham, are a joy. I am thankful, as well, for the love and support of Toni and David Owen and Jean Vivian over the years. In addition, I want to thank Lilly and Mo Zuberi, Anita and Sofia Zuberi, and Steve Chase for their kindness and for welcoming me as part of their family.

I am blessed to have two beautiful and wonderful daughters, Saskia and Naomi, who have given me so much joy and happiness. They have opened up my heart in ways I never imagined. I also thank them for napping, so I could write every day!

Finally, my greatest thanks go to my partner, Dan Zuberi, whose love, respect, and friendship gave me the strength and confidence to complete this work. As my best friend, intellectual companion, and wonderful father to our two children, he has been the constant source of strength and joy in my life. This project would not have been possible without his love and support.

Introduction: Anti–Water Privatization Movements in the Age of Globalization

Stockton: "Let the People Vote!"

On the evening of February 19, 2003, a crowd of people gathered inside the city hall in Stockton, California, to observe a city council vote on whether or not to approve a private water contract for the city's water services. The atmosphere was emotional and intense as dozens of activists went head-to-head with business leaders and politicians and spoke out against water privatization. Despite the boisterous protests of anti–water privatization activists, proponents of the private sector model were firm that Stockton's water would not be at risk under a private contract. After more than two hours of heated debate, the crowd grew silent as the council members cast their votes; in a vote of four to three, the private contract was approved. This decision marked the culmination of a two-year battle between the mayor and his supporters and a coalition of activists determined to prevent the privatization of the city's water.[1]

The fight began in the spring of 2002, when the popular conservative mayor, backed by a majority on the city council, announced plans to privatize the municipal wastewater treatment plant, through a public-private partnership (P3) model. The mayor and his supporters on the council argued that contracting with a private company to run the city's water system would save the city money and keep water rates low. Soon after the initial plans to privatize the water system were announced, a coalition of environmental, labor, and voter rights organizations began to mobilize in opposition to the contract, organizing regular delegations to council meetings to speak out against the proposed privatization deal. Many of the people involved in the anti–water privatization movement were unhappy with the mayor's right-wing agenda and were already involved in political campaigns opposing his reelection. This group of individuals included Bruce Owen, a retired businessman and board member of a local chapter of a national environmental organization and one of the founding members of the coalition steering committee.[2] "The mayor and council effectively shut us out that night," he told me as he described the February 19 city council meeting. He called the opposition to privatization a "natural fight" against the mayor and his "ultra conservative, anti-government" position.

The coalition in Stockton soon began holding meetings to organize a campaign against the privatization of water services. According to Bruce Owen, the strategy was to present a "rational and reasonable argument to the mayor." "Our first priority was to gather the facts," he said, explaining that the coalition wanted "to make sure that we had a valid analysis to present to council." As part of their strategy to appear "rational" and "professional," some members of the steering committee opposed the recommendation by other coalition members—union members and youth activists, in particular—to use more militant disruptive tactics, including street protests and sit-ins. Owen felt that the "radical" element of the coalition and their focus on "candlelight vigils" would detract from the professional approach of the steering committee, and result in the dismissal of their arguments by the mayor and council. "We needed to be professional and business-minded," he said, adding that the coalition chose to focus on pushing for a public referendum on the privatization issue rather than staging protests because they "didn't want to be seen as too radical or as working for the unions."

Despite a concerted effort by a coalition of union members, environmentalists, and voter rights advocates to block privatization plans by the city of Stockton, divisions within the movement about the choice of tactics and frames—including the decision to focus on a ballot initiative—failed to generate widespread mobilization and also divided members of the coalition. Many of the union members who were involved in early mobilization efforts felt alienated by the focus on voter rights rather than on the risks from corporate control of water. While the coalition steering committee emphasized local democratic process, the plant employees—who feared losing their jobs under a private contract—felt it was important to focus on the negative track record of multinational water firms in terms of job losses and water quality. Still other members of the coalition stressed the global nature of the problem and wanted to draw attention to the negative consequences of water privatization in other communities around the world as well as the risks to local democracy from international trade agreements.

Bruce Owen and others on the steering committee believed that these arguments would shift attention away from what they believed was

the critical issue: the democratic accountability of the municipal council. Owen thought that heavy involvement by the union representing the
plant workers would be a "conflict of interest." Although he appreciated
their support for the cause, he was convinced that the coalition should
remain neutral and not overtly support the union in their efforts to safeguard their jobs. He also believed that a focus on the global nature of
the problem would detract from the local nature of the struggle and the
importance of focusing on municipal electoral politics. He described how
he felt the conflict between the more radical elements of the coalition
and those who advocated a less disruptive tactical approach ultimately
prevented the movement from blocking privatization because it allowed
the city a significant head start. "We failed because we started too late,"
he explained. "By the time we got organized to do the initiative, the city
was already making plans in secret in smoke-filled back rooms that the
public was not aware of. Meanwhile, we were busy holding vigils. That
hurt our cause."

In the fall of 2002, when it appeared that the Stockton City Council would proceed with the plans to privatize the water system despite
growing opposition from the public, the anti–water privatization coalition organized a ballot initiative that would require voter approval on
all privatization contracts. They began organizing and collecting signatures, and by February 2003, had gathered over eighteen thousand signatures, successfully qualifying the initiative for a public vote on the issue
of privatization.[3]

In light of the predicted ballot victory for opponents of privatization, the Stockton City Council decided to accelerate the vote on the
private water contract. On February 19, 2003, just thirteen days before
the scheduled ballot initiative vote, the city council voted to approve the
$600 million contract, handing full control of the city's drinking water,
sewage treatment, and storm water systems to OMI-Thames Water—one
of the major multinational corporations that own and operate water services around the world. Two weeks later, on March 4, 2003, the ballot
initiative organized by the anti–water privatization coalition passed by a
margin of 60 percent. Yet, it was not retroactive, and thus did not reverse
the city council's decision to privatize Stockton's water treatment system.[4]

The members of the coalition were furious and defiant, declaring that
the battle would continue. During the next thirty days, they attempted to
gather enough signatures to overturn the city council vote in a referendum, as allowed under California law.[5] Despite their efforts, and as a result of a counter-referendum campaign headed by the mayor of Stockton

and OMI-Thames, the coalition was unable to collect the required number of signatures for the referendum to be placed on the ballot and to repeal the vote by the city council. The anti–water privatization movement was unable to stop the privatization of the water treatment plant; ownership of the Stockton water system was turned over to OMI-Thames in March 2003.[6]

The failure to prevent privatization was devastating, with major consequences for water treatment and delivery in Stockton, including job losses at the treatment plant, increased water rates, and lack of investment in the required facility upgrades by OMI-Thames. Yet, rather than giving up in the face of defeat, the anti–water privatization coalition—along with prominent environmental and voter rights organizations—filed a lawsuit under the California Environmental Quality Act (CEQA) in May 2003. With the goal of reversing privatization of the treatment plant, they charged the city of Stockton and OMI-Thames for failing to complete an environmental assessment required under California law for any major new construction or facility upgrade.[7]

The court battle lasted for over five years, with judges twice ruling in favor of the coalition. Both of these decisions were appealed by the city of Stockton and OMI-Thames, which were granted a new trial in 2004. After losing a third appeal in July 2007, the city of Stockton decided not to appeal the decision by the California Superior Court in favor of the coalition members. As a result, they voted to rescind the contract with OMI-Thames, a move that cost the city $1.5 million. After five years under the control of a private company, the Stockton wastewater treatment plant was returned to municipal control in 2008.[8] Despite the initial loss of the anti–water privatization movement, their successful ballot initiative, combined with the council's tactical error in rushing the decision to privatize, set the stage for future victory. After a long and costly legal battle, the anti–water privatization coalition ultimately prevailed over the city of Stockton and succeeded in overturning the private contract.

Vancouver: The Globalization of Risk

On a warm June evening in 2001, several hundred people attended a public meeting in Burnaby, British Columbia, to voice their opposition to the Metro Vancouver region's plan to privatize the Seymour water filtration plant. As municipal bureaucrats and elected officials waited inside the theater to begin their presentation to the public, a group of activists made their way from the nearby subway station along the street toward the theater. People chanted slogans and sang, while drums beat out a constant rhythm that grew louder as the crowd neared the entrance. Others carried banners with such slogans as "Keep our water public!" and "Don't P3 in our water!" A ripple of blue—a theater group, dressed in flowing blue costumes with faces painted blue and silver—snaked its way along the street. As they danced silently into the theater and surrounded the elected representatives, they resembled a wave of water, flowing and moving in unison. "Those dancers were amazing," said Amanda Jones, one of the organizers of the anti–water privatization protest, as she described the scene. "I remember them dancing behind the chairman of the meeting, doing all these crazy movements right behind him with their costumes and their banner, and it was like his head was just going to spin off his neck. I mean the imagery was so fantastic. It was brilliant." The meeting was the beginning of an intense public battle between the Metro Vancouver government and anti–water privatization activists over the decision to contract the region's water services to a private firm.

The decision to outsource water services occurred in the spring of 2001 when the Metro Vancouver Board announced plans to contract a private sector company to design, build, and operate a water filtration plant at the Seymour reservoir in North Vancouver, which supplies water to 40 percent of the region's population.[9] Because of the high cost of the proposed treatment plant—estimated at $150 million over twenty years—the board made a decision to privatize the construction of the plant in order to reduce the financial burden for taxpayers. The proposed contract was the largest public-private partnership contract ever proposed in Canada.[10]

The announcement of plans to privatize the water treatment plant sparked a public outcry and within days a coalition of concerned citizens and organizations had formed in opposition to Metro Vancouver's proposal. Prominent environmental, labor, and social justice organizations organized public information sessions, sent delegates to the meetings of the water board, and commissioned a legal opinion demonstrating the risk to local control over water from multinational trade agreements, such as NAFTA.

Amanda Jones, a community activist and thirty-seven-year-old mother of two children, was one of the leaders of the anti-privatization movement. She had been involved in international human rights and social justice campaigns since she was a teenager. In 2001, she was working as a community organizer for a national social justice organization, the Citizens Action League (CAL), when she heard about the Metro Vancouver proposal to privatize the construction and operation of the new water filtration plant. The proposed privatization plans came at a time when she had been working closely with local, national, and global water activists to organize a major international water conference in Vancouver to be held later that year. She had also recently returned from Cochabamba, Bolivia, where she learned about the negative consequences of water privatization firsthand, including enormous rate increases and cut-offs in poor neighborhoods that had sparked a mass uprising, ultimately leading to the cancellation of the private water contract. These experiences shaped her concerns about water privatization, and her belief that "water is life" and should be protected as part of the commons.

Amanda Jones had only recently been hired by CAL, and soon realized this was her first major campaign. She was excited and eager to use her background in activism to help mobilize a major opposition movement to water privatization. She and her colleagues quickly began compiling information about water privatization, including stories about deleterious consequences from communities around the world and the negative track records of many of the major multinational water corporations that were bidding for control of the region's water. Because CAL had a national water campaign and was involved globally in movements to protect water as part of the public domain, Jones could draw on the resources and social networks from these national and transnational movements. At the same time, with her background in community activism, she was also able to tap into a broad-based network of activists and organizations on the ground in Vancouver and the surrounding communities. She described the importance of working in coalition with other community organizations:

How activism in general worked in Vancouver meant that we usually did most of our campaign organizing in coalition with others. And so we called a number of different organizations together that would have some kind of interest or expertise in the issue, like unions and environmentalists and antipoverty groups, to see what we could do collectively. And we found people were really interested in the issue. It was kind of a confluence of interests and timing that brought people together to get involved.

Mobilizing wide public support was considered vital to the campaign and CAL's vast and diverse networks were instrumental in bringing together a broad range of organizations with a stake in keeping water services in public hands, including environmental groups and labor unions. Many of the activists from these organizations had recently collaborated on anti-globalization and anti-trade campaigns and were able to turn to these existing networks to mobilize a broad-based community coalition against water privatization.

Drawing on tactics, opportunities, and narratives from the broader anti-globalization movement, activists in Vancouver emphasized the threat to local democracy, accountability, and community control from global economic institutions to mobilize public opposition to privatization and reframe the way decision makers viewed the issue. They also used disruptive and creative tactics designed to draw attention to the importance of water as a source of life. "We put out flyers and did media work, but we also went out in costume on buses and the subway with information to try to bring people to the meetings," explained Amanda Jones when describing the tactics used to mobilize the public and raise awareness of the dangers of water privatization. "We were also strategic in what we did at public meetings. We planted people around the audience that were prepared to ask specific questions and we had some chants prepared and that kind of thing." Activists objected to the control the politicians exerted and, as she explained, "decided that we really needed to take over the meetings because they [the Metro Vancouver Board] had sat us all down like a bunch of students, totally contrary to popular education style, and *they* were going to tell us what the scoop was. You know, all of that attitude. So we felt we needed to direct the meeting from our point of view and let them know that we understood the issue perhaps better than they did. And we did that really successfully, and it shocked them." In Vancouver, the anti–water privatization activists had a well-planned and organized campaign to disrupt meetings and counter the pro-privatization arguments presented by the Metro Vancouver Board.

Faced with pressure from the coalition of activists and municipal leaders, the Metro Vancouver Board agreed to hold three public information sessions in the region. At the first meeting, in Burnaby, more than five hundred people packed a crowded hall, with many more turned away due to fire regulations. Those who attended represented a wide range of environmental, labor, and social justice groups. There were also representatives from the majority of the municipalities in the Metro Vancouver region. The meeting lasted over four hours, as hundreds of people opposed to water privatization lined up to speak at the microphones and repeatedly requested a reversal of the board's decision to privatize the water filtration plant.[11]

Two weeks later, more than eight hundred people turned out at the second public meeting in North Vancouver, with the overwhelming majority speaking out against privatization. The meeting was scheduled to last two hours, but continued well into the night, finally concluding after five hours. Similar to the previous meeting, Metro Vancouver representatives faced a hostile audience, who regularly shouted angry boos at the bureaucrats speaking from the podium in front of the crowd. Many of the movement leaders spoke, as did members of the general public—united in their opposition to the privatization of the water filtration plant, with most of the concern centering on the impact of international trade agreements on the ability of the regional government to regulate and control water resources.[12]

While the meetings were noisy and disruptive, anti–water privatization activists drew on expertise demonstrating the risks from international trade agreements on the local control of resources and strategically presented a unified message around trade and economic globalization. CAL was instrumental in bridging the concerns of diverse actors and organizations under the common rubric of threats from international financial and trade treaties. "What we were able to do working with other groups was to show how trade and privatization concerns were important for environmental issues and jobs and poverty issues. Everything just sort of came together around those concerns. And you know that was really effective because in the end, that is what convinced Metro Vancouver. They were really worried about losing control under NAFTA," Amanda Jones explained.

The third meeting never took place. Faced with the enormous public turnout in opposition to water privatization at the two previous meetings and the overwhelming negative public sentiment, the Metro Vancouver Board ultimately reversed their decision to contract out water services to a private water firm, citing concern over trade agreements and their effect on local governments' control over resources. In a press release issued that day, they revealed the rationale for the decision; the board's chair said, "[T]he public who attended the public consultation meetings, and those who corresponded with us, sent a very clear message. There is uncertainty about the impact of international trade treaties, NAFTA and GATS, and even though the risks may be small, the public did not want to take those risks no matter what efficiencies may be gained. We said we would listen. And we did. We took that 'sober second look' and changed our minds."[13]

Anti–water privatization activists in Vancouver drew on preexisting networks, tactics, and frames to build a movement against water privatization, pointing to the risks from trade agreements to local democratic accountability and the capacity of municipal governments to regulate and protect environmental resources. This strategy was instrumental in facilitating widespread mobilization and opened up opportunities at the political level for a favorable outcome. In a brief but fierce campaign, the anti–water privatization coalition in Vancouver had succeeded in preventing the privatization of water and keeping water services under public control.

Comparing Social Movements Against Water Privatization

The stories of Bruce Owen and Amanda Jones, two activists involved in the anti–water privatization movements in their communities—Stockton, California, and Vancouver, British Columbia—reflect the widespread opposition to water privatization that continues to play out in many communities around the world. Yet the stories from these two activists also demonstrate that, despite responding to similar threats, movements against water privatization evolve differently. Why did these movements take different forms? What are the factors that shape the success of local movements responding to municipal governments that propose to outsource services to multinational corporations? What are the implications of the outcomes of these movements for local counter-globalization movements more generally? This study answers these questions by examining the mechanisms and processes that affect movements resisting neoliberal globalization on the ground. My aim is not simply to identify the conditions that explain differences across similar movements, but also to illuminate the pathways by which these types of movements can both successfully resist global corporate hegemony and shape social policy at the community level. While the power of international economic institutions is increasing, the forms of resistance at the local level offer hope for creating alternatives to economic globalization, in part because they have clear targets and channels for participatory democracy. The stories of anti-water privatization activists in Stockton and Vancouver demonstrate that "one-size fits all" globalization is not inevitable, and that alternative visions are made possible through the power of social movements and the strengthening of local democracy.

The Globalization of Protest

Social movement scholars have increasingly focused on the effect of economic globalization—often referred to as "globalization from above"[14]—on mobilization, including the emergence of new transnational movements that challenge the hegemonic power of global institutions. Globalization, scholars argue, is changing the nature of protest by forging new collective identities and frames, creating new transnational networks of activists, transforming the relationship of movements to the state, and shifting targets from domestic to global institutions.[15]

For example, Jackie Smith's research on the 1999 anti-WTO protests in Seattle found that global forces have created new transnational movement actors whose targets lie beyond the domestic level. Activists involved in the "battle of Seattle" targeted global institutions, used frames that reflect the global nature of the struggle, formed cross-national and cross-movement networks for cooperation, and adopted new collective identities that reflect a shared understanding of globalization.[16] Other scholars argue that globalization transforms collective identity through the growing integration of civil society. The research by Donatella della Porta and her colleagues on the 2001 G8 protests in Genoa and the 2002 European Social Forum in Florence, for example, demonstrates how globalization transforms people's everyday lived experiences and creates new identity bonds. While global activism does not eradicate national or territorial identities, they argue, activists involved in these protests consider themselves part of a global civil society that responds to transnational issues.[17]

Research on the globalization of protest points to the growing power of international institutions, and suggests that scholars must look beyond the state to the increasingly powerful "global polity" in which it is embedded, in order to understand how it changes the dynamics of domestic social movements, altering the relationship of movements to the state.[18] For example, in their book *Activists beyond Borders: Advocacy Networks in International Politics,* Margaret Keck and Katherine Sikkink examine the role of transnational advocacy networks in shaping the global policy field. Challenging the state-centered conceptualization of power, they highlight the agency of non-state actors and point to the growing power of individuals and social movements that contest economic globalization and the shift in power from the nation-state to the international realm. Networks of activists and organizations—linked together through a global policy arena—pool resources, share information, and join forces to target offending governments, in what Keck and Sikkink call the "boomerang effect." In this model, domestic actors—whose grievances are ignored by their own governments—appeal to transnational advocacy networks of international organizations that, in turn, target their own governments to put pressure on the aberrant government. These transnational advocacy networks are able to influence domestic states and other governments internationally, through "persuasion, socialization and pressure."[19] In illuminating the complex processes of interaction and negotiation between states and non-state actors, Keck and Sikkink demonstrate the multiple sites of power and pathways to change in an increasingly globalized world.

At the same time, while many social movement scholars point to the emergence of new transnational movements and networks in response to globalization, some maintain that globalization is limited in its capacity to shape domestic movements because it cannot provide the resources, networks, and opportunities necessary to sustain mobilization in the way that domestic factors can.[20] For example, in his book *The New Transnational Activism*, Sidney Tarrow argues that what some call new transnational movements are more likely the temporary cross-border organization around specific and limited issues rather than sustained social movements. He contends that the outcomes of these movements as well as their ability to sustain mobilization are still dependent on the political opportunities found in the particularities of the nation-state and the strength of domestic social networks and resources. While activists may be connected to global networks and engage in protest at the global level, he maintains, they remain "rooted" in their specific localities and the "social networks . . . resources, experiences, and opportunities that place provides them."[21] While globalization has created new opportunities, targets, and spaces for activists, local issues remain critically important for sustaining mobilization.

Yet, few studies focus on anti-globalization movements at the local level. While there has been considerable attention paid to globalization and protest, most research focuses on movements that operate at the global scale, including transnational networks, campaigns, and events.[22] There has been a lack of research on the effect of global processes on local movement organizing. Transnational policies and institutions have an impact at the local level. New opportunities and threats shape the way local movements mobilize by providing a new source of grievance, new targets for action, and new alliances.[23] There is a need to distinguish between the dynamics of local versus transnational movements resisting globalization to understand the context-dependent effects of transnational forces.

Global capital, multinational corporate power, and international trade and financial institutions have increasing influence over domestic governments and their ability to regulate and control local resources. At the same time these global business entities reshape local movements and the way they respond to opportunities and frame their grievances.[24] For example, Gould, Schnaiberg, and Weinberg examine local worker-community movements in the context of the globalization of the political economy. They argue that local movements operate within a "framework of rising transnational flows of liquid capital, commodities and corporate services," that create both new opportunities and constraints for local movements.[25] Domestic movements must adopt new models for mobilizing local activists, they contend, including forging alliances between local and transnational movements as well as building cross-sectoral coalitions of movement actors. In other words, local movements must seize both domestic and international opportunities to be successful in a globalized world.

Recognizing the multi-scalar nature of movements in a globalized world is particularly important for local environmental justice movements. While these movements address specific community-based environmental problems, their grievances and movement outcomes are increasingly shaped by decisions and power structures outside of the local community, including those at state, national, and international levels. For example, in their book *Chronicles from the Environmental Justice Frontline*, J. Timmons Roberts and Melissa Toffolon-Weiss examine local environmental justice movements in Louisiana and the connections between local environmental justice movements and national and international processes. They argue that "these struggles do not take place only at the local level"; what happens at the state, national, and international level also matters.[26] In a changing political economy—in which national and global institutions and corporations have increasing power over local governance and the economy—Roberts and Toffolon-Weiss demonstrate that environmental justice movements must build bridges between local community organizations and causes and national and international networks of environmental organizations. Decisions that impact the environmental health of local communities are often made outside the community, at the state, national and, increasingly, at the international level. Activists are thus more likely to have favorable outcomes if they build alliances—particularly cross-class and cross-race coalitions—that enable them to fight the battle for environmental justice at multiple levels.[27]

The multi-scalar nature of local politics and mobilizations is particularly important in the case of water privatization, in which multilevel institutional processes influence local decision making, and create new barriers and opportunities for mobilization. Many water scholars argue that water governance is becoming progressively more "glocalized" and point to the emergence of a new politics of water, in which the local and global are increasingly connected.[28] And while some scholars contend that economic globalization and the dematerialization of the economy has disembedded the economy from place, resulting in "the end of geography" (a point at which local context no longer matters to policy making), the conflict over resources, including life-sustaining resources such as water, demonstrates that the local context matters enormously to decision making and governance.[29] Joachim Blatter, Helen Ingram, and Pamela Doughman claim, for example, that the local has become more important with the reordering of the global economy and the growing power of global institutions. They argue that the decentralization of power from nation-states to municipalities has increased the strength of local communities and their voice in global affairs.[30] Local governments often have more power than national governments vis-à-vis international trade and investment treaties, for example, because they are not the principal signatories of these treaties and because social movement actors have more access to local policy makers and thus can shape local governance policies. Activists at the municipal level have the capacity to influence policy decisions and protect local resources from the encroachment of international institutions because they are able to participate in regulatory and decision-making bodies.[31]

Local anti–water privatization movements are well-suited for investigating how and why global forces are reshaping local movements because they operate at the global-local nexus. Externally, they are shaped by global forces, including the push for neoliberal reforms at the municipal level, the pressure from multinational corporations to outsource water services, and the impact of international trade and investment treaties. Multinational water firms, transnational policy networks, and global financial and trade institutions offer global solutions to what they perceive as a local problem—the decline in public investment for water infrastructure and services. At the same time, these movements are intensely local, drawing on local constituents for mobilization, and dependent on the dynamics and structures of municipal governments. While global institutions are indeed hegemonic, they are more vulnerable to countermovements at the local level because they represent clear and tangible targets for resistance, and because local political opportunity structures are more accessible to social movement actors than those in the national or international arena. As these movements bring together locally rooted networks and resources—often from diverse and previously divided movement sectors—in order to protect the well-being of the community, they are more socially cohesive and reflective of a shared collective identity than anti-globalization movements occurring in transnational spaces. Local movements that recognize the interplay between the global and the local are well positioned to counter the pressure of increasing encroachment of global capital on local environmental systems and resources and offer viable alternatives to the neoliberal project.[32]

In this book, I argue that what happens at the local level matters for the outcomes of globalization and for the regulation of environmental resources. I demonstrate that local anti–water privatization movements which link local and international concerns, as well as build coalitions with a wide range of organizations at both the local and global level, draw attention to the risks from multinational corporations, international regulatory organizations, and financial institutions and create local solidarity in the face of international threats. The research challenges the growing tendency of social movement scholars to conceptualize the responses of social movements to globalization as a scale shift: a shift away from domestic modes of organization and contention to transnational protest. In contrast, I argue that successful challenges and alternatives to neoliberal globalization will not necessarily come from movements operating at the transnational level, but rather from locally situated counter-hegemonic movements that are connected globally but rooted in local communities.

Comparing Social Movements Against Water Privatization: The Cases

In the Contested Water study, I focus on local social movement organizing against water privatization. From the perspective of social movement actors, I examine the reasons why these movements emerge in response to the outsourcing of water services. Through a comparative analysis of two anti–water privatization movements—in Stockton, California, and Vancouver, British Columbia—I examine the factors that shape the emergence, development, and outcomes of these movements, including how global processes shape mobilization on the ground. Table 0.1 presents the similarities and differences between the two cases. While activists in both places were facing similar threats—the proposed outsourcing of local water services to global water firms—and shared similar beliefs in the importance of protecting water as part of the commons, they responded differently. Both places involved cross-movement coalitions but used different frames and tactics, with the movement in Stockton focusing on local political process and relying on a voter-driven ballot initiative, and the movement in Vancouver focusing on mobilizing broad public support and drawing attention to the global risks from international institutions. In Stockton the water system was privatized for five years; it took until 2008 for the coalition's legal challenge to win and have the private contract overturned. The movement in Vancouver was immediately successful—the water system was not privatized. Despite the eventual convergence in terms of outcomes, the two movements clearly experienced divergent trajectories that reflect the complex movement-building conditions of each place. Getting to know these differences is key to understanding how similar threats play out across different contexts and for understanding the particularities of local social movement mobilization in an era of globalization.

Table 0.1
Comparing Stockton, California, and Vancouver, British Columbia

Cases	Stockton, California	Vancouver, British Columbia
Proposal	2002 proposal to outsource water services Multinational water firms short-listed for contract	2001 proposal to outsource water services Multinational water firms short-listed for contract
Response	Desire to protect water as part of the commons Cross-movement coalition organizes in opposition	Desire to protect water as part of the commons Cross-movement coalition organizes in opposition
Frames	Focus on local political process	Focus on global threats to local democracy
Tactics	Voter-driven ballot initiative Electoral referendum	Grassroots mobilization Legal opinion on NAFTA
Outcome	Privatization in 2003 Private contract overturned in 2008	Water remains under public control

Methodological Approach

To investigate the nature of anti–water privatization movements in North America, I chose two contrasting cases because of several factors. Stockton, California, and Vancouver, British Columbia, represent distinct local responses to similar global phenomena and comparing them reveals insights into what differentiates local responses from transnational processes. Both movements occurred at relatively similar times (between 2001 and 2003), and in the wider global context, took place during a time of heightened anti-globalization protest and the rise of transnational movements. This gave me the opportunity to examine the effect of global processes on local movements.

Both cases also share similarities beyond the global context of transnational protest. In both cities, the short-listed companies were all subsidiaries of major multinational water corporations, and in each case, the privatization contracts were the largest ever proposed in the history of each country and had gained significant national and even international attention. As a result, both places were considered as "must-wins" by the anti-privatization movements in order to stop the spread of water privatization to other municipalities. Both cases also involved broad community coalitions, including labor and environmental organizations. Yet while both movements were responding to similar forces and shared similar threats, they differed significantly in terms of form and outcomes, with the movement in Stockton initially failing to prevent privatization and activists in Vancouver successfully overturning the decision to privatize water. This variation in movement trajectories and outcomes provided the opportunity to empirically uncover what factors—including the role of global processes—enabled or constrained these two movements.

While Stockton and Vancouver differ in many ways, including municipal structure, population, and along socioeconomic lines, what makes them interesting and important to compare is that they number among the few cases globally in which mobilization against protest occurred prior to the privatization of the local water system, in contrast to other cases where mobilization occurred after water systems had already been privatized. The Stockton and Vancouver cases are also examples of movements occurring in communities in advanced industrialized countries and thus are more comparable with each other, than with other cases in developing or underdeveloped countries, such as Bolivia, Ghana, and South Africa. Many of the latter cases—the movement in Cochabamba in particular—have been examined in the literature on water privatization and governance; what has received less attention are the struggles against water privatization that have emerged in cities in the Global North.[33] While it is unsurprising that similar trajectories of resistance have occurred in communities in poorer countries—in part because grievances are often linked to the politics of distributive justice as well as to wider post-colonial struggles[34]—anti–water privatization activists have mobilized in cities across the Global North, ranging from Stockton and Vancouver to Atlanta and Berlin, often with the similar levels of intensity and mass mobilization. This book examines two such cases.

My research design was organized around this goal: to identify, through comparison, the mechanisms and processes that enable or constrain anti–water privatization movements and to understand how context matters for movements responding to similar threats. I completed seventy digitally recorded in-depth interviews with activists involved in the movements in Stockton and Vancouver from April 2008 to January 2009. The respondents in the two communities represented a diversity of organizations and institutions, including environmental, labor, social justice, community organizations, and government/political institutions.[35] A comparative analysis of the dynamics of coalition building from the perspective of movement actors highlights the similarities and differences that characterize mobilization processes and sheds light on context-dependent factors, including political-institutional and organizational culture and place-based understandings of water, as well as external factors, such as transnational political and economic processes. A comparative method is especially useful in examining complex patterns of interaction because it increases explanatory power through the identification of different pathways that lead to particular outcomes.[36]

I supplemented the in-depth interviews with extensive analysis of textual documents, such as media stories and organizational documents related to the two cases. Analysis of these textual sources allowed me to identify key issues around decision making, issue framing, tactics, and strategies for mobilization to complement the patterns obtained from interviews as well as to understand the differences in how issues are understood and framed by different groups across time.

The seventy semi-structured interviews lasted approximately one-and-a- half to two hours and consisted of both open-ended and survey questions.[37] The interviews allowed me to collect detailed information about people's participation in the movements to explicate factors that influence their participation, including their values and behaviors and contextually mediated understandings of water. The interviews also generated data on the role of frames, political opportunities, and coalition building and how these processes shaped movement trajectories and outcomes. Finally, the interviews enabled me to understand the role of global processes—including transnational opportunities, networks, and frames—in shaping the movements in Vancouver and Stockton. The data from the transcribed interviews, reflection notes. and scanned and downloaded documents were coded and patterns of responses identified using Qualitative Social Research's *NVivo 8* software.

The goal of the research has been to develop an understanding of local social movements for creating progressive social change in an era of uncertainty and complexity. In a rapidly changing world, where cultural and economic production shifts across national and international boundaries, comparative qualitative research helps illuminate the pathways by which these social transformations shape individual life experiences and sheds light on the relationship between institutional power and the subjective "on-the-ground" experiences of social movement actors.

While there are limitations to the methodological approach, including the small number of cases, the interviews with activists in Stockton and Vancouver allowed for an in-depth understanding of how similar mechanisms are shaped by the context of contention.[38] At the same time, the comparison facilitated the development of hypotheses about the factors that shape local mobilizations against water privatization—including the importance of local-global linkages—that are useful for understanding similar struggles in communities around the world. Anti–water privatization movements are linked together through a shared resistance to the commodification of water and a common desire for equity in water governance. As global water activist Vandana Shiva argues: "Water wars are global wars, with diverse cultures and ecosystems, sharing the universal ethic of water as an ecological necessity, pitted against a corporate culture of privatization, greed, and enclosures of the water commons."[39]

Overview of the Book

My comparative study of the anti–water privatization movements in Stockton and Vancouver uses a process-oriented approach to social movements, and examines the dynamic interaction of mechanisms of contention, including ideology and frames, opportunities, and networks to explain the similarities and differences across the two cases. From the analysis of these two divergent cases, important patterns emerge that advance the understanding of localized forms of contention in the broader context of globalization.

Chapter 1 focuses on the wider political and economic context of water privatization, water governance reform, and anti–water privatization movements. I discuss the evolution of municipal water systems, the rise of neoliberal globalization, and the changing role of the state, as well as the emergence of multinational water companies and the growing trajectories of resistance that are surfacing in response to water privatization. I argue that anti–water privatization movements are an understudied and important area of research for explaining the conflict over resources—including life-sustaining resources such as water—and for illuminating the pathways to equitable and just water governance policies.

Chapter 2 develops a theoretical and methodological framework for understanding local forms of resistance to globalization that enables a more holistic, multilevel, and dynamic conceptualization of counter-globalization movements at the local level, including the interplay between ideology, frames and networks, and global and local institutions. I suggest a way of moving beyond the static, top-down approach to the study of contention and consider the ways in which movements to protect the commons—in particular, water—can integrate and advance theories of environmental sociology and social movements.

Chapter 3 examines the context-dependent meanings of water that shape political mobilization by activists in Stockton and Vancouver. I specifically examine how constructions of water are tied to situated knowledge and are influenced by notions of power, justice, and spirituality. These symbolic meanings not only affect our everyday lived experiences but also explain why many of us have a strong desire to mobilize against the privatization of water in our communities. This chapter demonstrates that social movements—particularly those that seek to protect the commons—are shaped by more than structural mechanisms such as political opportunities and networks and are also influenced by the social and cultural experiences of movement actors, including their emotional and visceral attachments to place.

Chapter 4 focuses on differences in the mobilization frames between the anti–water privatization movements in Stockton and Vancouver. Here, I move beyond an examination of broader ideological beliefs to investigate the role of strategically constructed frames in shaping the diverse trajectories and outcomes of the two movements. I demonstrate that linking global and local frames is critical for strengthening the claims of local movements responding to global processes.

Chapter 5 focuses on differences in political context and opportunities between the two cases and the impact of these differences on movement trajectories and outcomes. I consider the role of multilevel opportunity structures—at the local and global scale—for understanding the dynamics of local movements in a global context. I make the case that movements which synthesize global and local opportunities generate a sense of local solidarity, uniting authorities with activists in a shared desire to protect local resources from global economic and institutional threats.

Chapter 6 examines cross-movement coalition building in the anti–water privatization movements in Stockton and Vancouver, and describes how movements to protect water as a public resource facilitate alliances between traditionally oppositional sectors, including labor and environmental organizations. This chapter reveals that in the context of localized resistance to globalization, alliances between the labor and environmental movements—united under a common frame and shared tactics—can be a potentially powerful force against the increasing commodification of nature.

Chapter 7 contributes to debates about local social movements in a globalizing world, and advances our understanding of the dynamic processes that combine to shape mobilization, from an on-the-ground perspective. This chapter also adds to a sociological understanding of the role of social movements for resisting and offering alternatives to economic globalization. It also presents some of the implications for similar movements, with recommendations for how to build successful local campaigns. My research on anti–water privatization movements reveals the growing link between local contestations and global power structures. While most research on globalization emphasizes the hegemony of global capitalism, my research reveals the counter-hegemonic power of communities, and demonstrates that the best opportunities for challenging globalization will not necessarily originate from transnational movements, but rather from local movements that are connected to global networks and resources.

1

From Commons to Commodity: Water Governance, Neoliberalism, and the Privatization of Municipal Water Systems

Water may be the single most critical element of life—it is the largest and most complex habitat on Earth, nourishes every species on the planet, and is the primary material of all living things—but we don't have a clue about its true nature.
—Jeffrey Rothfeder, *Every Drop for Sale*

Water is unique as a natural resource. No human can survive without access to water. At the same time, increasing demand and environmental threats—including industrial pollution, agriculture, urbanization, overconsumption, and climate change—have given rise to a global water crisis, with fresh water resources disappearing at an alarming rate. Scarcity makes water more economically valuable while also intensifying conflict and competition.[1] Much of Africa, Australia, the American Southwest, and the Middle East are currently facing serious issues of scarcity and conflict over access to fresh water. As water levels and quality decline, demand for water rises and the world's capacity to meet the needs of current and future generations is endangered. Today, over 1.1 billion people do not have regular access to fresh water, while over 2.5 billion lack access to sanitation services.[2]

The massive increase in urbanization worldwide has put tremendous pressure on municipal water systems to provide both clean drinking water and sanitation services to the billions of people in need. In many cities, lack of infrastructure and high levels of poverty prevent access to clean drinking water or sanitation services for billions of people around the world. In urban areas globally, continued disinvestment in municipal infrastructure has left cities grappling with how to pay for critical upgrades to water service infrastructure.[3] The question of how to provide the world's population with clean drinking water and sanitation services has become one of the most important and contested challenges of the current century.[4]

In the face of this global crisis, water is increasingly being recognized as a fundamental human right in order to address issues of inequity and as a way to compel governments and policy makers around the world to guarantee every citizen access to an adequate supply of clean water.[5] Yet, despite the recognition of water as a basic human right that is necessary for survival, the world's water crisis continues to worsen, as increasing numbers of people lack access to critical life-sustaining water resources.[6] How do we understand the escalating water crisis and the conflicts that emerge from it? Beyond its social and ecological dimensions—including threats from climate change, degradation of water systems from pollution and overuse, and population growth—the water crisis is first and foremost a problem of governance.[7] In order to understand the scope of the problem it is thus necessary to examine the historical roots of water governance and regulation.

This chapter focuses on the wider political and economic context of water privatization and water governance. First, the chapter discusses the transformation in water governance that has occurred over the last thirty years, including the changing role of the state, the influence of global institutions and water firms, and the effect of neoliberal globalization on water service restructuring and management. Next, the chapter examines the scholarly literature on water privatization and water governance and the growing recognition of the importance of equity and justice to water management. The final section of the chapter discusses the rise of anti–water privatization movements globally and reflects on the importance of examining these movements to understand the conflict over water at the local level as well as shed light on the role of civil society in shaping water governance policies.

The Politics of Water: Governance Reform, Neoliberalism, and the Changing Role of the State

Water has become one of the most critical policy fields of the twenty-first century. Over the last thirty years, governments, policy makers, financial institutions, and civil society groups at local, regional, national, and international levels have increasingly recognized that beyond issues of scarcity, pollution, and investment, the global water crisis has emerged and is exacerbated largely as a result of mismanagement of water systems and poor governance.[8] Attention to the crisis intensified in the 1990s with the organization of several key international conferences that highlighted problems of water scarcity and pollution, along with the staggering number of people around the world with little or no access to clean drinking water.[9] Out of these conferences emerged an international consensus that major reforms are needed in the areas of water governance, including stronger regulation and management strategies, to deal with water scarcity and degradation and to address issues of conflict around water.[10]

In 2003, the United Nations went so far as to identify water governance failure and mismanagement as the cause of the world's water crisis.[11] The recognition of the primacy of governance in alleviating the world's water crisis has heralded a new era of water governance reform. New global water policies—enshrined in international documents such as the "Dublin Statement on Water and Sustainable Development," adopted by the United Nations in 1992[12]—call for global water governance models that emphasize greater public participation, the importance of local knowledge and experience, increased efficiency, and a focus on equity and sustainability in the distribution of water resources.[13]

The push for governance reform has been particularly strong in urban areas globally, as cities grapple with problems of water scarcity, degradation of water systems from overuse, industrial pollution, and runoff from agricultural fertilizers and pesticides.[14] Pressure on urban governments to develop new water policies and management regimes is also the result of stronger environmental regulation, increased regional conflicts over water supplies, and a downshift in regulatory oversight and stewardship from state or federal governments to local communities.[15] Decentralized decision making, coupled with disinvestment in public services and infrastructure, has left cities with few resources to build, expand, or upgrade water utilities, at a time when water infrastructure is deteriorating.[16] The result is an environmental and social water crisis as local communities struggle to provide an adequate supply of clean, high quality drinking water to their populations and develop policies to address issues of scarcity and manage water resources more efficiently.[17] This problem is exacerbated in countries of the Global South, where problems of drought, poverty, and inequality present further obstacles to the implementation of water governance policies that are ecologically and socially just. In many parts of the world, uneven service delivery between urban and rural areas and between economically well-off and poorer populations creates further challenges for governments seeking to increase efficiencies and expand infrastructure to reach vulnerable and marginalized populations.[18]

How do local utilities adapt to these new challenges? Governments and policy makers have begun to recognize that new models of management and governance that focus on local policy making are required to deal with both ecological and social imperatives of the water crisis as well as the need to adapt to changes in the institutional field of governance and investment. In the United States, for example, Megan Mullin argues that ecological threats, stronger environmental stewardship policies, aging infrastructure, and a growing demand for scarce resources from urban sprawl and population growth have led to a major shift in water governance and a "new local politics of water" that recognizes the importance of local governments.[19] Municipalities are focusing more and more on governance reform to deal with the escalating water crisis. One of the biggest transformations in water governance policies, she contends, is the shift in focus from large-scale water projects, financed by the federal government, to small-scale decentralized public policies emphasizing water conservation and efficiency, resulting from growing public opposition to mega-projects and disinvestment in infrastructure funding by the federal government. The new local politics of water is one in which, Mullin argues, "governance has replaced technology in the new era of public water supply management and local decisions are paramount."[20]

While Mullin correctly points to the devolution of power from federal to local governments as shaping the current water policy field, downshifting is only part of the story. Beyond domestic political and ecological dynamics, local restructuring and governance reform is also shaped by broader global economic and political transformations that have occurred over the last three decades, including the rise of transnational institutions and networks and the rise of neoliberal globalization.

Since the 1990s a shift in power dynamics from domestic to global institutions has transformed state-market relations, affecting the provision of goods and services, including water services, in communities around the world. With the rise of economic globalization, many scholars argue that decisions related to service delivery and resource management are increasingly influenced by transnational institutions and regulatory bodies.[21] For example, Bronwen Morgan points to the emergence of a new global water policy field that has shifted power over water governance and management toward the transnational level. She argues that this reordering of power is a result of four major institutional changes that have decentered the state, including the emergence of global water organizations and water-specific conferences, the growth of transnational water firms, the rise and entrenchment of international financial and regulatory bodies, and the appearance of transnational water policy networks. These new transnational institutions and networks of non-state actors have growing influence over domestic governments and how water services are delivered in communities around the world.[22] A key example is the World Water Council, which holds its international conference—the World Water Forum—every three years, and is made up of representatives of business-based nongovernmental organizations (NGOs) and large multinational corporations that have considerable influence on global water policies. As Morgan argues, "[E]ach World Water Forum hosts a formal intergovernmental ministerial meeting whose declarations are scrutinized by key actors in the policy field for clues to transnational trends, making the Forum a powerful site of non-state actor influence on global water policy trajectories."[23] The result has been a push for market-based delivery of water services along with the growing tendency to view water as a commercial commodity.[24]

Other international agencies include the Global Water Partnership, and the World Commission on Water, which include top-level representatives from the World Bank, the International Monetary Fund, and most major multinational water companies. These agencies recognize water as an economic good and actively promote water privatization as a way to increase efficiency, ensure full-cost recovery and expand water service delivery.[25] Transnational water firms, supported by global financial institutions and a network of international water agencies that promote private sector participation in water treatment and delivery, lobby and pressure municipal governments to outsource public water services.[26]

While countries in the Global South, including Africa, Asia, and Latin America, have seen the highest growth areas for water service restructuring over the last two decades, water privatization is not restricted to the developing world. It has become a truly global phenomenon. With the spread of neoliberalism in the 1990s, and the withdrawal of the state from the delivery of public services, water marketization has increased across Europe and North America, particularly in England, Germany, and France, and in the last decade, in the United States and Canada.[27]

Many scholars argue that the entrenchment of powerful financial institutions and the growing privatization of water services across the globe is linked to the rise of neoliberalism, where deregulation, decreased public spending, and the decline in public sector services have resulted in the growing shift of capital into new social and ecological domains.[28] Neoliberalism—often called economic globalization or simply globalization—emerged in the 1970s with the rise of Margaret Thatcher in the United Kingdom and Ronald Reagan in the United States and is characterized by a rolling back of the Keynesian welfare state and a shift to market-oriented economic policies.[29] These policies—often referred to as the "Washington Consensus"—include deregulation and privatization of public sector institutions and services, trade liberalization, decreased public spending, erosion of labor and employment protections, and an emphasis on individual liberty over the collective social good.[30] The promotion and expansion of neoliberal globalization is linked to the increasing power and influence of global institutions such as the World Bank and International Monetary Fund and the expansion of international trade agreements.[31]

As capital becomes increasingly mobile and national and local governments are more beholden to international trade agreements and the pressures from global economic integration, private sector involvement in the delivery of public services is rapidly growing.[32] Municipal infrastructure has become a particular target for restructuring, as deteriorating facilities combined with severe budget constraints (a problem likely to become worse as a result of the current global financial crisis and the adoption of strong economic austerity measures) encourage governments to turn to the private sector for financial investment in both infrastructure and services. Multinational corporations also target municipal governments for lucrative and relatively risk-free infrastructure contracts, because local governments are often unable to resist the pressures from both corporate lobbyists and federal governments pushing private sector investment.[33]

Water has become a focus of economic restructuring as multinational water corporations compete for ownership of the world's fresh water. New global regulations in trade and services have emerged, along with increased pressure by international financial institutions to make water privatization a condition for loans to developing countries[34] Because water—and, by extension, wastewater treatment—is so vitally important to human health, the economy, and the environment, it is a vulnerable target for private sector investment. Scarcity and necessity mean that investing in water services has become highly profitable for global capital investment, with many analysts referring to water as the "new oil."[35] Proponents of market-based models consider commodification of water the solution to the problem of water scarcity and declining infrastructure; the basic argument being that by assigning property rights to water, private corporations will invest in infrastructure, resources will be reallocated properly, and conservation will be encouraged.[36]

Yet water privatization was not advanced solely as a means of increasing investments in infrastructure and ensuring more efficient use of water resources. While some argue that water privatization was promoted as a lucrative investment for the private sector and a way for governments to allocate resources more efficiently, more recently private sector water delivery has been endorsed and encouraged by the World Bank and other global financial institutions, as a way of correcting the failures of public water management, including increasing environmental conservation, reducing social and economic inequities—particularly in developing countries—and providing clean water to the billions of people who currently lack access worldwide.[37]

Since the 1980s, public water systems have been under attack by proponents of market-based models of governance, who argue that "governments are less productive, efficient, and effective than markets."[38] With the massive governance failure on the part of the public sector, including poor planning, crumbling infrastructure, and pollution and degradation of water systems, private sector involvement in water systems has been hailed as a means of correcting those failures and increasing equity in terms of access to and affordability of water services by poor and marginalized populations.[39]

Development policies in particular, including those adopted by the World Bank, emphasized the importance of investing in sanitation and water services as a means of increasing productivity of communities around the world, alleviating poverty, and improving environmental conservation of water systems.[40] With the ideological shift toward neoliberalism and the adoption of the Dublin Principles—and its recognition of water as an economic good—the World Bank's support for water privatization increased from the 1990s onward. The idea was that the private sector would introduce key measures to improve water governance and the management of water utilities, including increased efficiency, full-cost recovery, and investment in infrastructure upgrades and expansion.[41] The result has been a dramatic increase in private investment in urban water systems globally. For example, in 2000, more than 460 million people—the majority located in the Global South—had their water supplied by a small number of transnational water firms, most of them European. This number is expected to grow to over 1.5 billion by 2015.[42]

Governance Reform and Water Privatization: Where Are We Now?

What has been the outcome of the water privatization experiment over the last three decades in cities across the globe? Despite the promotion of water privatization as a means of improving infrastructure and water quality, ensuring environmental sustainability, and achieving economic efficiency, the actual outcomes of outsourcing water services to private firms has been problematic and disappointing in many communities around the world, including cities in the Global North and South. Problems include rate increases, cut-offs for those who cannot afford to pay, and failure to complete promised infrastructure investments.[43] While many of these cities were ill-served by public sector water utilities prior to privatization, suffering from poor water quality, crumbling infrastructure, and lack of access for poorer residents, private sector involvement has proved to be no panacea. For example, in the South American cities of Cochabamba, Bolivia, and Buenes Aires, Argentina, privatization of water services failed to raise water quality standards or expand infrastructure.[44] In Cochabamba, water rates rose significantly—for some their monthly bill doubled or even tripled, forcing them to spend more than half their monthly earnings on water—effectively cutting off poor residents from accessing clean water and sanitation services.[45] Similar problems have occurred elsewhere. In the United Kingdom, for example, the privatization of local water utilities under the Thatcher administration was a massive failure; poor management, lack of investment in infrastructure, skyrocketing water rates, and deterioration of water quality forced the government to intervene and strictly regulate private water firms.[46]

While reducing inequities in access and distribution is often a major goal in water governance restructuring that involves private sector investment, privatizing water often fails to address issues of poverty and inequality, which frequently lie at the root of inequities in the distribution of and access to clean water and sanitation services for vulnerable populations in many parts of the world.[47] Many scholars argue that this is a result of private sector approaches that focus heavily on market-based concerns such as full-cost pricing and increasing efficiencies at the expense of addressing social and political inequality.[48] Other scholars argue that the rise of water privatization in the developing world is a form of neocolonialism as water firms based in the Global North gain control over an increasingly large portion of the world's water supplies, with more and more people in the Global South dependent on European or American corporations for access to the most crucial life-sustaining resource.[49]

A further critique stems from the narrow economic paradigm that considers water first and foremost a commodity to be regulated by the market, which prevents the recognition of other critical factors that should be considered when developing water policies, such as cultural understandings of water and historical land rights.[50] Even when such issues are taken into consideration, they are often trumped by market concerns, placing the economic value of water over other issues including social and ecological equity. Rutgerd Boelens, David Getches, and Armando Guevara-Gil, for example, argue that while neoliberal water restructuring policies often explicitly recognize local understandings of water and indigenous culture and rights, these cultural and equity considerations are often reconstituted to fit within the ideological framework of water privatization and, as a result, do not take precedence over market rationality and pricing.[51]

In communities around the world, the unexpected and damaging results have raised local resistance to water privatization as well as increased government regulation to monitor and oversee private water service delivery.[52] Critics of water privatization, including those who mobilize in their communities to prevent or overturn the privatization of local water utilities, emphasize the importance of a new model of water governance and decision making that, while not excluding the importance of valuing water, must take into account other fundamental issues, such as environmental and social equity. A future of water determined by market forces and not by equity concerns, they argue, will exacerbate the world's water crisis and continue to prevent billions of people around the world from access to life-sustaining water resources.

Politicizing Water: The Importance of Social, Political, and Ecological Equity

Water has always been a hotly contested issue, but recently, with the expansion of water privatization and the growing tendency to view water as a commercial commodity, the debate over the future of water has become one of the most important policy issues. Many scholars argue that because water is a life-sustaining resource that has enormous social, cultural, political, and economic value, it represents a key starting point for understanding the limits of market-based governance models.[53] These scholars contend that water marketization fails to account for the multiple meanings and values of water and thus call for new ways of thinking about water and policies that integrate the multiple dimensions of water, including its ecological, social, economic, and political aspects.[54] John Whiteley, Helen Ingram, and Richard Perry argue that problems in governance are linked to a failure to recognize the multiple values that are associated with water and suggest that governance reform must "recognize the multiple and incommensurable meanings of water in all of its specific geographical and historical sites of encounter."[55] Without a new model of water governance, the water crisis will worsen, and conflict over access to water resources will continue.[56]

One of the main critiques of water privatization and efficiency models is that they focus too narrowly on pricing water as a way to deal with the water crisis, and thus fail to recognize the importance of other critical elements of water supply and service delivery, including well-managed and effective institutions and social and environmental equity.[57] Many water scholars argue that conceptualizing water as a commodity both simplifies it and disconnects it from its multiple and complex dimensions, such as its ecological, social, cultural, and spiritual values and properties.[58]

Beyond commodification, other scholars point to the limits of current policy frameworks for integrating noninstrumental values of water into decision making. For example, Blatter, Ingram, and Doughman point to the "inadequacy of standard legal, engineering, and economic frameworks for capturing value-laden and symbolic attributes with which water is endowed."[59] They argue that instrumental rationality and the reliance on science-based methodologies alone ignores or downplays nonmaterial meanings and values of water, what they call "cultures of nature." As a result, water privatization has largely failed to increase equity in terms of distribution, affordability, and access, especially for vulnerable and marginalized populations. They contend that integrated approaches to water governance are necessary for developing policies that are sustainable, equitable, and just, because such policies recognize the full complexity of water resources in their social and ecological contexts.[60]

What does an integrative approach look like? There is wide consensus among water scholars that to deal with the global water crisis, a new model of governance must first and foremost address broader notions of fairness, equity, and justice. A model based on these principles will compel policy actors to focus on marginalized or vulnerable populations and consider issues of social and political power in the processes of negotiation and decision making.[61]

While much of the governance reform that has occurred over the last thirty years includes some recognition of the need to take an integrated approach to water governance and consider the cultural and social dimensions of water, reforms have had little success increasing access and affordability of water resources to vulnerable populations. Scholars argue that this is in part due to the difficulties of implementing efficiency policies in the face of existing on-the-ground conditions, such as institutional failure, political corruption, poverty, and inequality, as well as the legacy of colonialism and imperialism.[62] For example, in her book *Privatizing Water: Governance Failure and the World's Urban Water Crisis*, geographer Karen Bakker demonstrates how the political and social context of inequality can act as a barrier to effective governance reform. She argues that in many countries around the world, including Bolivia, Argentina, South Africa, and Ghana, a history of structural inequalities stemming from a legacy of colonialism and racialized policies—with the harsh conditions for nonwhites during the apartheid regime in South Africa being particularly pernicious—have prevented poor and vulnerable populations from benefiting from new decentralized water governance policies designed to distribute water resources more evenly and fairly. Governance policies based on economic efficiency versus equity, she argues, cannot overcome these barriers because they ignore the social dimension of water resources. As a result, even policies that are designed in part to improve access and water quality for poor populations are unable to achieve these outcomes without addressing historical and structural inequalities.[63]

Margaret Wilder also emphasizes the need to include equity considerations in global water governance reform. Her research on water governance reform in the state of Sonora in northern Mexico demonstrates that while water governance reforms were successful in increasing political equity by decentralizing decision making and incorporating local participation into their framework, these reforms were unable to increase economic equity and expand water infrastructure and services to rural or poor populations. She argues that this is partly a result of global governance models that emphasize economic efficiency over social equity, but also because neoliberal policies—including an end to subsidies for poor residents and rising water rates—have increased levels of poverty and inequality in the region.[64] She contends that the "gains in political equity represented by codified, formal opportunities to participate in water policy decisions have been more than offset by adverse changes in economic equity related to liberalized markets, loss of subsidies and consumer-pays water pricing principles."[65] Rising inequality and the lack of political and economic power can act as powerful barriers for disadvantaged populations to participate equally in the policy arena, and, as a result, policies tend to favor political elites, wealthy residents, and corporate interests.

Much of the research literature on water governance points to the lack of emphasis on social equity considerations for the failure to improve distribution of water services. In outlining these failures, scholars are increasingly making the case that to solve the world's water crisis policy reform must address poverty and social inequality, especially when implementing policies and practices to increase water access and allocation among poorer populations in developing countries. When social dimensions of water governance are given equal weight to economic and efficiency concerns, they claim, the focus becomes how to provide water services to those who lack access or who cannot afford to pay for clean drinking water and sanitation.[66] Governance models based on the principle of equity would rectify the failure of current reforms to ensure fair access to life-sustaining water resources, because they explicitly recognize social and political inequalities and thus the goal becomes distributive justice of water resources.[67]

How do we integrate equity into global water governance policies? Much of the literature on water governance suggests that water must ultimately become a political issue and thus a question of democracy, accountability, and institutional access. In this light, most water scholars agree on the importance of certain key political reforms, including decentralized decision making, and inclusive public participation, especially by poor or marginalized populations, and the use of legal tools to rectify institutional failures related to land and water rights.[68] A decentralized approach allows for public participation in water policy decisions and water management and the recognition of the needs and values of local context and culture.

While much of the governance reform advocated by the major water players around the world, including the World Bank, the World Water Council, and the United Nations, emphasizes the importance of decentralization and public participation, critics of current reforms insist that most participatory models are flawed because they are not representative of the broader population and thus policy outcomes tend to favor powerful groups; those with greater access to material and intellectual resources. In order for community-based participation to be effective, critics argue, it is essential to avoid "elite capture" by ensuring that the voices of those less well-off, including poor, marginalized, or vulnerable populations, are heard.[69] Water policies, with the goal of achieving equity in distribution and access, must include a formal role for marginalized populations to ensure that local participation in the management of water resources is "more than just the formal trappings of participation."[70] Beyond equity, scholars contend that local knowledge of water resources and successful stewardship practices are critical for developing policies that work on the ground.[71] A democratic and accessible policy arena is fundamental for understanding value pluralism, avoiding conflict over the competing demands on local water resources, and ensuring that policy outcomes achieve equity in accessibility, affordability, and productivity of water.[72]

While many water scholars point to the importance of achieving social and political equity in water governance, Karen Bakker adds another dimension to the equity framework that recognizes the unique nature of water resources. She argues that beyond sociopolitical notions of equity, ecological equity must also take center stage in the global water policy arena because, unlike most other public services, water is directly linked to geography. Policy reforms must therefore recognize the ecological dimension of water in order to protect and conserve scarce water resources. She argues that we need to expand the current model of distributive justice from one that focuses on the redistribution of material resources to one that allows us to "act collectively as stewards of the socialecological lifeworlds of which we are a part."[73] Ensuring that ecological equity considerations play a central role in water governance models, Bakker argues, will shift the debate over water governance from the public-private dualism that currently characterizes the global water policy field to "ecologically sensitive systems of governance," that focus on conservation, stewardship, and community solidarity.[74] The central concern of an ecocentric model is not whether water is a public or private resource, but rather whether the interests of the environment are balanced with those of other users. A focus on ecological equity would thus allow for effective private sector involvement with strong regulatory oversight by public governance institutions.[75]

While the call for a new model of governance based on equity is widespread, the traditional efficiency framework continues to dominate the water policy field with water privatization on the rise in communities around the world. The result of this narrow model of governance is the ongoing conflict over water resources. Protest movements against the commodification of water continue to mobilize at the local level and increasingly at the international level, with the rise of a global movement for water rights. What has also become abundantly clear over last three decades is that water—a life-sustaining and increasingly scarce resource—has been fundamentally politicized.

The Politics of Water: Mobilizing Against Water Privatization

The call for a new politics of water originated in the local contestations that have emerged in response to water governance reform, which has dominated the global water policy field since the 1980s. When local populations perceive water management policies to be inequitable or not reflective of the multiple meanings of water, conflict emerges, ranging from protest to political instability. In communities around the world, clashes between local governments and citizens' movements around the management of public services and resources have become widespread. Privatization of water services and growing inequity in access to and affordability of water services has spurred widespread global resistance movements—from Cochabamba, Bolivia, to Orange Farm, South Africa, to Atlanta, Georgia—that challenge the logic of commodification and the deregulation of the global economy at the expense of social and environmental well-being.[76]

These movements are diverse and are often made up of cross-sectoral coalitions, including environmental groups, labor unions, and social justice organizations. Many are connected to broader anti-globalization struggles. With the rise and entrenchment of neoliberal globalization, activists are increasingly concerned about the risks to local resources—particularly life-sustaining resources such as water—and seek to prevent losing local control over resources in light of the growing power of transnational institutions.[77] In other cases, including anti–water privatization struggles in Bolivia, Argentina, and Indonesia, "water privatization was linked with broader concerns about macroeconomic management, the behavior of political classes, and the appropriate role of the state (and foreign capital) in the national economy."[78] Activists involved in these struggles oppose the status of water as a commercial commodity and focus on the social, cultural, and ecological dimensions of water. For them, the logic of the market does not apply; access to water is a human right and part of the global commons and therefore should not be treated as a commodity. These activists advocate water policies that reflect the principles of ecological, social, and distributive justice in order to ensure the conservation and fair and equal distribution of water for current and future generations.[79] The goals of these movements reflect the model of equity that is advocated by water scholars.

Many of these struggles are linked to the failure of local and state governments to ensure equity of access, distribution, and water rates under a private system. For example, in 1999, in order to deal with governance corruption, inequalities in access to water services, and crumbling infrastructure, Bolivia's federal government signed a forty-year contract with Aguas del Tunari—a subsidiary of the U.S. company Bechtel—to deliver water services in Cochabamba, the third largest city in Bolivia. After privatization, water rates dramatically increased to levels that many local citizens could not afford, resulting in their water supply being cut off.[80]

In response, a group of activists, led by labor unions and environmentalists, mobilized the public to protest the rate increases for water services. These protests culminated in a general strike, which shut down the local economy for four days.[81] Unable to reach an agreement with either the private company or the government, the activists continued to protest in the streets, occupying the downtown core. The protests were declared illegal by the government, which sent in hundreds of troops to disperse the protesters. Violence ensued, and after several days of unrest, thousands of people were injured, many severely, and several people were killed.[82] As a result of massive public pressure and fear of increased backlash and violence, the Bolivian government canceled the contract with Aguas del Tunari, returning responsibility for the management and delivery of water services in Cochabamba to the public sector.[83] Although the movement in Cochabamba was reacting to the unjust policies from privatization, including rate hikes and service cut-offs, the response was as much about the abdication of responsibility on the part of the government to ensure that its citizens could access water services at an affordable price.[84]

While resistance to water privatization has increased since the 1990s, along with a growing global movement to declare water a human right, these contestations are rarely the main focus in literature on water governance. The increasing pressure on governments to outsource water services and the resistance to privatization by anti–water privatization movements presents a valuable research opportunity for examining the dynamics of counter-hegemonic resistance. While many scholars argue that counter-globalization movements should be examined to gain a better understanding of globalization and its alternatives[85], few studies specifically investigate the concrete social processes by which this counter-hegemony occurs, including the mobilization of resources and organizational networks by social movements on the ground in response to global forces. In the case of water privatization, while there has been considerable research examining the policies and results of outsourcing water services, there has been little research examining the countermovements that arise in response to water privatization in local communities.

To understand the complexity of the global water crisis as well as how to achieve equitable and just solutions to supply water to global populations, it is critical to examine the trajectories of resistance against water privatization. While much of the scholarly literature on the global water crisis has dealt with key issues of water governance reform, including efficiency, rights, identity, and equity, so far the examination of anti-privatization movements has been superficial, particularly from the perspective of social actors who are involved in these movements. This book addresses this weakness and offers a unique and important contribution to the understanding of water privatization and the conflict between commodification and the commons. It focuses on the growing struggle against the privatization of water services in two communities, and tells the story from the ground up; from the perspective of the activists who organize in their communities. It sheds light on the nuances, complexities, and pathways of local resistance to global economic transformations. As the first scholarly examination of the social processes underlying movements against water privatization, it provides a unique lens onto some of the most important and dynamic struggles of the current century as communities around the world struggle to deal with issues of water scarcity and provide clean and affordable water services to their populations.

This chapter has focused on the broader context of water privatization and water politics, including governance reform and the new institutional orderings at the global level as well as the challenges to current efficiency models and the rise of conflict over water privatization in communities around the world. The question of water governance is one of the most important policy arenas of this century, and its importance will only increase. Enormous investments are needed to fix crumbling water systems across the globe and expand services to populations that lack access to clean water. The expansion of private sector investment in water service delivery, coupled with the considerable influence of transnational water institutions in shaping global water policies, will spur new movements that resist commodification and confront the new politics of water. The intimacy and connectedness of water to people's everyday lives, their daily needs, their health and well-being, and, in some cases, their very survival, mean that the groups mobilizing to protect water are among the most powerful and visible movements of our time. Regardless of one's stance on private sector investment in water services, investigating the dynamics of global water movements is critical for advancing the understanding of social change, and developing equitable and sustainable water governance policies. If the global water crisis is not solved—and in a way that is just and equitable—the risks to ecological sustainability and social well-being, as well as to international political and economic stability, will be severe, if not irreversible.

2

Globalization and Local Social Movements: Understanding the Dynamics

How do we understand the processes that underlie counter-globalization movements? How have sociologists and other scholars approached the analysis of these movements, specifically local mobilizations focused on protecting environmental resources? Much of the current research on globalization and protest examines the shift in contention from the domestic to the transnational realm. Yet, local movements have emerged in response to perceived threats from globalization in many communities; while these movements are shaped by international processes, they are dependent on and rooted in locally based networks and institutions.

Understanding how globalization shapes mobilization from an on-the-ground perspective is particularly important for explaining movements that seek to protect environmental resources because of the complex interaction of locally bounded resources and place-based attachments to nature with international institutions. Investigating anti–water privatization movements offers a potential avenue to help clarify current theoretical debates about the effects of globalization on localized forms of contention because these movements are shaped by both global and local processes. My findings from the anti–water privatization movements in Stockton and Vancouver reveal that global processes do not play out uniformly across different places, but rather interact with context-dependent factors including political institutions, organizational networks, and socially constructed meanings of water.

In Stockton and Vancouver, anti–water privatization activists faced similar external threats and expressed strong and deeply held emotional attachments to water. At the same time, the movements used remarkably different frames to present their claims and fight the privatization of their water services. The Stockton organizers kept their focus narrow and local, and actively resisted those within the movement who wanted to expand the frame by linking the argument to broader global issues. In Vancouver, movement organizers had previous experience building broad coalitions and quickly seized on the use of global frames, with immediate effect.

What explains the differential development of these movements in response to the similar threat of water privatization? In this chapter, I describe a conceptual and theoretical framework to explain these differences. Building on the work of scholars in the area of social movements, the environment, and globalization, I focus on the interactions between ideology, frames, political opportunities, and movement coalitions, and their role in shaping the mobilization patterns of the two movements in the context of local responses to neoliberal globalization and policies related to environmental resources.

Local movements implicated in transnational flows of capital and power should be studied differently from movements operating at a global scale because the processes that shape them operate in different ways at the local versus the national or transnational level. Global processes—including flows of capital and international institutions such as multinational corporations and international trade and investment treaties—provide both new threats and opportunities for social movements at the local level, affecting the kinds of networks, frames, and tactics used by activists. In the case of ecological resources, globalization has reordered the nature of environmental politics by shifting power from local communities to international institutions and corporations. At the same time, natural resources—water in particular—remain geographically bound and thus inextricably linked to the politics of place.

How do we understand the struggle around resources and environmental risk in light of these changing power dynamics? Local counter-globalization movements offer a way to clarify theoretical contestations around the political ecology of natural resources, including the multi-scalar nature of environmental movements and the relationship between local political struggles and powerful transnational forces.

Research on Social Movements

Over the past fifty years, there has been increasing interest in the study of social movements and the development of a theory of contention that explains not only why, but how, individuals engage in activism. Scholarly research in this area generally falls into one of three categories: *frame analysis*—which investigates how movement organizations construct meanings so as to mobilize constituents and influence policy decisions; *political process*—which examines the role of the political institutional context in shaping mobilization and outcomes; and *network analysis*—which examines the importance of individual, organizational, and movement ties for mobilization and outcomes. Over the last ten years, scholars of contentious politics have highlighted the need for a more dynamic approach to the study of contention that focuses on the interplay between these three overarching categories.[1] While these three areas of investigation offer valuable theoretical tools for understanding and explaining social movements, there are limitations to each approach, particularly for understanding the dynamics of local movements resisting neoliberal globalization and seeking to protect environmental resources.

A growing body of research investigates the increasing importance of globalization and the role of transnational institutions, frames, and networks for shaping social movements. Many scholars argue that transnational institutions have growing influence over the distribution of resources and environmental risk and claim that social problems, in particular those that are ecological in nature, are increasingly disembedded from local context, affecting the way local movements mobilize.[2] Yet environmental sociologists have identified the importance of local context for shaping constructions of nature and mobilization around environmental resources.[3] Research on water resources and water privatization also underscores the emotional and spiritual attachments that people have to water and water resources; values that are deeply associated with situated knowledge, including people's everyday lived experiences. Locally rooted noninstrumental values and meanings of water shape mobilization around water resources.[4]

Environmental social movements—particularly those that seek to protect the commons—are shaped by more than structural mechanisms such as political institutions and networks. They are also influenced by the social and cultural experiences of movement actors, including their emotional and visceral attachments to place and how they view their role in creating social change. The multi-scalar and multi-mechanistic nature of local counter-globalization movements makes it clear that understanding how globalization influences environmental social movements on the ground—including how it shapes ideology, frames, network dynamics, and political opportunities—requires a new, more dynamic model of contentious politics.

A Dynamic Approach to Contention

Examining the global-local nexus in relation to environmental resources and policies requires a broadening of our understanding of social movements and social change. Globalization generates dynamic and complex transformations of social, political, and economic institutions as well as how individuals view the world around them, which must be explained. Many scholars argue that understanding how societies and nature are reconstituted by global flows of capital, networks, and resources requires a more fluid sociological analysis of global complexity.[5] John Urry, for example, makes the case that globalization and localization are intertwined through flows of resources and networks, and suggests that scholars move beyond traditional concepts of social structures and societies, and pay more attention to these "mobilities" and their interdependence.[6]

Some scholars, including Jeff Goodwin and James Jasper, have criticized the overly static and structural nature of traditional theories of contention and proposed a more dynamic model for investigating the complex processes of social movements. Goodwin and Jasper argue that the political process model restricts the analysis of contention to an examination of static political structures and thus minimizes agency and emotion, as well as the relational and cultural dimensions of contention.[7] Research on non-state processes that shape mobilization highlights the importance of social-psychological, cultural, and relational mechanisms such as identity, meaning construction, and social networks, as well as how they combine together and influence movement emergence and outcomes.[8]

Social movement actors do more than respond to and engage political opportunities; they create and open new opportunities through social actions and constructed meanings. Activists have the capacity to alter and create opportunities through frames, choice of targets, social ties and alliances with elites, and through preexisting movements.[9] Some prominent social movement scholars have addressed the overly structural, static nature of some conceptions of contention and have called for a more dynamic approach to social movement theory that recognizes the complex interplay between different mechanisms and processes of contention.[10] In *Dynamics of Contention*, for example, Doug McAdam, Sidney Tarrow, and Charles Tilly construct a model for understanding complex episodes of contention. They argue that previous models are limited because they do not recognize the dynamic, interactive, or recursive features of contentious politics, and treat them as static or linear. Examining mechanisms, such as resources, political opportunities, networks, and frames as discrete variables, rather than processes operating in tandem to produce episodes of contention limits a full accounting of the causal conditions of protest. In light of these limitations, they call for a dynamic model that recognizes the processes and mechanisms that combine in varying historical contexts. They stress the need to focus on *environmental* (external conditions of contention such as resources and opportunities), *cognitive* (individual and collective interpretations and perceptions including ideology and frames) and *relational* (interpersonal or interorganizational connections) mechanisms that work together to form broader episodes of contention operating in a similar fashion across different situations.[11]

This dynamic approach to the study of contention—including an examination of the relationship between ideology and frames, political opportunity structures, and networks—is useful for understanding the interplay between global processes and local movement dynamics. Global processes shape different aspects of movements, altering collective action frames, providing new opportunities for mobilization, and creating new coalitions and networks. A model that allows for an investigation of the intersection of these processes and mechanisms is critical for understanding social movement complexity in a globalizing world. The mechanisms that shape local movements in the context of globalization are neither linear nor static, but rather are interactive, continual, and recursive.[12] Local counter-globalization movements are shaped by the interaction of global and local political and economic opportunity structures as well as by the pre-movement ideological beliefs of social movement actors.

A dynamic model allows for a full explanation of the stages of social movements including their emergence, development, and outcomes. Yet few studies use a dynamic approach to contention. Most research on social movements focuses on the emergence and mobilization stages of movements. Some scholars have argued that this is because research on social movements often privileges one causal mechanism and its impact on mobilization, such as political opportunities, rather than the multiple and complex processes that shape movements across their full cycles, including outcomes.[13] The broader, more dynamic analysis of contention presented in this book focuses on the interplay between internal processes, such as individual and organizational ideology and frames, and external processes, including multilevel political and economic opportunity structures and social networks.[14]

The Meaning of Nature: Ideology and Environmental Activism

A reworking of the current model of contentious politics is particularly important for explaining mobilization around environmental resources—especially the struggle over life-sustaining resources such as water—because of the complex and multiple sociocultural understandings of nature that influence political activism around the environment. Beyond institutional structures, networks, and resources, these meanings of nature add a critical cognitive dimension to the dynamics of environmental politics. Understanding how environmental activists view the world around them—including the meanings they attach to nature and how they perceive their role in challenging corporate and institutional power—is essential for explaining mobilization around environmental resources and the possibilities for alternatives to neoliberal globalization.

Traditionally, social movement scholars have downplayed the cognitive dimension of movements in favor of institutional structures, resource mobilization, and network dynamics.[15] While there has been increased attention on the discursive strategies of social movements, previous research has generally focused on mobilization frames—the collective messages and narratives movement organizations use to make their claims— rather than on individual pre-movement ideology or understandings of the problem. Some scholars have pointed to the importance of distinguishing between these two cognitive mechanisms in order to understand the distinct ways that they shape mobilization at different stages of the protest cycle.[16] This distinction is of particular importance for explaining environmental movements because of the emotional and personal attachments that individuals have with nature. While movement frames are critical for understanding the collective meanings of a particular movement, because they are by nature *collective*, frames alone cannot explain the full range of cognitive mechanisms, including the ideological and social-psychological bases of mobilization.

Social movement scholars have also identified the importance of examining ideology as a key mechanism for explaining why people engage in activism. Pamela Oliver and Hank Johnston, for example, point out that most research on cognitive work in social movements tends to amalgamate the concepts of ideology or broad belief systems with the strategic marketing of ideas—or frames—into one overarching process of contention. They argue that this conflation has obscured a rigorous theoretical understanding of the separate role each process plays in shaping social movements, particularly at different stages of contention.[17] Social-psychological processes of contention—such as emotions and ideology—should be examined separately from the more concrete and strategic work of collective action frames so that their role throughout the cycle of protest can be clarified.[18]

One of the best ways to refine our understanding of these distinct cognitive processes and how they are shaped by local context is to use a comparative approach. Social movement actors from similar movements in different contexts may reveal similar ideational understandings of a problem or phenomena that may motivate them to join a movement, and yet these understandings are translated into different frames of contention later in the movement cycle, resulting in divergent trajectories and outcomes. Clarifying the role of ideology and frames in social movements helps illuminate how intellectual and emotional understandings of problems—especially those that are ecological in nature—are used to mobilize collective action and construct strategic and instrumental frames of contention.

Studying ideology as an important and distinct process from frame construction can also help explain why and how opposition to a particular problem emerges *before* mobilization and the strategic construction of grievance frames occurs. For example, Rachel Schurman and William Munro's research examining the anti–genetic engineering movement demonstrates that the "thinking" work by scientists and intellectuals prior to mobilization was critical in creating the groundwork for a mass movement to emerge by articulating the problem and identifying the target. They demonstrate that social movements often begin with the intellectual work of a small network of individuals, and that without this pre-mobilization cognitive process, "many movements simply would never materialize."[19] They call for additional empirical research to understand how ideology is transformed into knowledge constructed for political action.[20]

Ideology is particularly important to understand when studying environmental social movements. Socio-natural relationships are critical to shaping our worldviews, and thus can shed light on why some people mobilize to protect these resources.[21] The research literature on water privatization and water governance points to the centrality of ideological processes for explaining the mobilization around water resources and the contested nature of local water policies. As described in chapter 1, research on water governance highlights the complexities and nuances around the management of water resources. Failure to recognize the multiple dimensions of water in governance planning—including its social, economic, environmental, cultural, and spiritual aspects—has made water a highly contested resource and shaped mobilization around water resources.[22]

Water is deeply imbued with social and cultural meanings and is an integral part of people's sense of place. Notions of identity, equity, respect, spirituality, belonging, and territorial and social rights are intricately tied to understandings of water and water governance.[23] As values around water are often intrinsic and not utilitarian, people's decisions, choices, and mobilization around water cannot be fully explained by instrumental rationality or political institutional context.[24] We need to understand social constructions of water as well as the identities and values of social actors in order to clarify how ideology shapes people's understanding of risk and their motivation to become politically engaged in movements to protect water resources. As geographer Karen Bakker argues, these multiple and complex dimensions of water integrate our social, spiritual, ecological, and political selves and explain the connection between individual values and political action.[25]

A Synthesis of Environmental Sociology and Social Movements

Why do people mobilize to protect environmental resources such as water? One way to clarify the ideological dynamics of mobilization around the environment—including water—is to draw on the work of environmental sociologists, political scientists, and geographers whose research sheds light on the transformation in socio-natural relationships over the last fifty years.[26] In recent decades, these scholars have found that the rise of environmental values has transformed the way individuals think about and interact with the environment. The dematerialization of the economy and the rise of global environmental risk—from pollution and deforestation to drought and climate change—have led to an ontological shift in how people view the world around them. Ronald Ingelhart, for example, argues that as countries become more prosperous, and their populations enjoy increased economic and social security, there is a corresponding shift from materialist values—emphasizing security—to postmaterialist values centered on quality of life and individual expression, including preservation of the environment, freedom of speech, and gender equality.[27] While anthropocentric notions of human domination over nature prevailed historically, an understanding of the interdependence of humans and nature shapes current conceptions about the environment.[28] This fundamental value shift in worldview and self-conception is important for explaining pre-mobilization attitudes and values that influence people's decisions to participate in broader environmental movements.[29] These social-psychological or ideological understandings of nature and resources differ from the strategic frames used by environmental movement organizations and should be examined as separate processes.

Theories of environmental sociology—particularly those that focus on the social construction of nature—provide a theoretical understanding of the ideological processes from which environmental movements emerge. Sociological research on the environment focuses on the socially constructed meanings that we attribute to the world around us as well as the social processes by which environmental conditions are constructed and recognized as problems. Cultural and ideological processes are particularly relevant for understanding environmental movements because of the complex relationships individuals have with nature and ecological resources. Beyond social movement frames, people draw on cultural symbols or metaphors to make sense of the world. We are constantly re-defining ourselves in view of technological changes and paradigm shifts in relation to nature. Theories of environmental sociology demonstrate that meaning is not derived by the nature of the material or external world, but rather in the social, political, and cultural contexts in which nature is constructed.[30]

Understanding how we assign meaning to our environments is essential for explaining the growing conflict over resources. Cultural differences in perceptions of risk and trust produce different meanings of nature across time and space. A good example is Terre Satterfield's research on the conflict over old-growth forests in Oregon, which examines our relationships to nature and how our interpretations of the world around us are shaped by local context. Satterfield uses a cultural analysis of the dispute between environmentalists and loggers over logging to understand how the processes of contention are expressed culturally. By examining locally situated practices, she is able to demonstrate that the imagined worlds of those involved—the social construction of nature and identity—shape the political discourse of the movement.[31] Research on environmental conflict reveals that our day-to-day interactions with our surrounding environment are critical for shaping our worldviews and for explaining both the consolidation and contestation of hegemonic ideas and practices.[32]

Drawing on theories of environmental sociology enhances the understanding of anti-water privatization movements by illuminating how social movement actors construct meanings of nature and environmental risk through the social, geographical, and cultural processes in which they are embedded. In the case of water, this embeddedness is particularly important because of the deep emotional attachments people have to water as the source of life.[33] Water has multiple spheres of meaning—from notions of a human right and a gift of nature to spiritual and community-building values—that shape conflict and competition and resistance to utilitarian governance policies.[34] The "locality" of water (both politically in terms of regulation and control and geographically through watershed boundaries) also provides the specific context for the problem that shapes the ideology of movement actors and helps explain how intellectual knowledge is translated into the strategic political knowledge that drives mobilization. Illuminating how people view and understand water helps clarify the difference between pre-mobilization emotional and ideological processes and the more strategic and instrumental frames used by movements to make claims and create opportunities for success.

The process of negotiating symbols and meanings of nature must be a central focus for studying environmental movements. These ideological processes shed light on pre-mobilization values and beliefs of social movement actors and help explain individual participation in broader movements focused on environmental change. Understanding how nature is embedded in our senses and through spatial and temporal boundaries can shed light on differences in metaphors of nature across cultures and how those differences explain variation in environmental policies, construction of risk and harm, and opportunities for mobilization across societies.[35]

Bridging environmental sociology with social movement theories benefits both subfields. While empirical studies of environmental social movements add rigor to theoretical explanations of the environment and socionatural relationships, environmental sociology—the social construction of nature in particular—helps heed the call by some social movement scholars to treat ideology as distinct from collective action frames.[36]

Framing Protest: Collective Meanings and Social Movements

Another key cognitive dimension of mobilization is the construction of collective action frames by social movement actors to make claims, mobilize the public, and target authorities. While ideological factors shed light on pre-movement values and attitudes, frames occur at later stages of the protest cycle, and are the *collective* meanings created by social movement organizations. As a way of moving beyond structural or static explanations of contention, many social movement scholars have pointed to the role of *frames*—the construction of meaning—in shaping collective behavior. Political opportunities are interpreted through cultural and cognitive processes, and thus the defining of opportunity is as important to movement mobilization as the existence of opportunity itself.[37]

Drawing on Goffman's idea of frame analysis—the ways in which we use cognitive schema to interpret the world around us—social movement scholars have emphasized the importance of interpreting opportunities and constructing grievances for mobilizing support for a movement's goals.[38] One of the most important framing strategies used by social movement actors is the creation of common messages that link together ideologically similar, but previously disconnected groups to mobilize broader support. Social movement organizations use "master frames" that resonate broadly, allowing them to move beyond single-issue, ideological-driven causes and build networks across diverse groups.[39]

Social movement scholars have also identified the importance of the interplay between frames and other key processes of contention, including political opportunities, organizational culture, and outcomes. Social movement organizations are more successful in drawing attention to their grievances and creating the conditions for outcome success when they construct frames that resonate with political opportunity structures.[40] For example, in their study of the American welfare rights movements, Ellen Reese and Garnett Newcombe found that, while organizations construct frames reflecting their core values and beliefs, those that are more pragmatic in their ideological outlook are often more successful in maximizing support for their cause because of their ability to construct frames that resonate with both authorities and the broader public.[41]

At the local level, the outcomes of counter-globalization movements are contingent upon the ability of social movement organizations to overcome ideological constraints and strategically link frames with local political opportunity structures. By synthesizing global and local frames, organizations can draw attention to the threats to local democracy from transnational institutions and bring together disparate networks by building a sense of local solidarity in the face of global foes.

Linking Local and Global Frames

Global political and economic processes have reshaped the way local movements mobilize, including whom they target and how they frame issues. Theories of risk and modernity point to the increasingly global nature of social and environmental problems, and the growing influence of global institutions over the regulation and distribution of resources and risk. Globalization has reordered the international political economy and disembedded ecological problems from a local context to a complex global political-economic system.[42] Governing ecological resources and addressing the consequences of environmental harm is increasingly difficult for local governments as control over resources shifts from domestic political institutions to international agencies and corporations. In many cases, the challenges for communities go beyond the complexities of a globalized political economy, to the unequal distribution of environmental harm. As political scientist Peter Dauvergne argues in *The Shadows of Consumption*, the expansion of the global economic system has shifted the ecological consequences of global consumption from wealthy communities to poor and vulnerable populations, who lack the power and resources to cope with increased pollution and environmental harm.[43]

At the same time, the context for social and political action has moved to the international arena. Social movements face new challenges dealing with transnational forces as targets shift beyond the geographic and political boundaries of the local community.[44] The globalization of environmental risk, through the transformation of power from domestic institutions and communities to international bodies and corporations, has been a catalyst for local resistance movements. Activists—particularly those involved in movements to protect water as part of the commons—fear a loss of control over local, life-sustaining resources and seek to prevent their commodification. These movements are increasingly adopting frames that reflect the global nature of the environmental risk.[45]

Social, economic, and environmental transformations above the level of the state reshape the way social movements operate at the local level since they must respond to global shifts. Local movements react to global problems with new ways of organizing and new frames that help activists understand the complexities of globalization and how they redefine local political context.[46] Increasingly, for local counter-globalization movements to be successful, they must target institutions beyond the local context and highlight the global nature of the problem. For example, Marcos Ancelovici's research on the French anti-globalization movement demonstrates that framing local issues as global problems helps social movement actors and organizations connect domestic concerns with events and processes that are exogenous to the state. He argues that "collective interpretive processes play an important role in explaining the dynamics of contemporary contentious politics because it is through them that actors make sense of long-term structural changes such as globalization."[47]

While studies on the use of global frames by local movements have illuminated the importance of bridging global and local issues, more in-depth research is needed to clarify how global frames are integrated into local movements and how they shape mobilization outcomes. Anti–water privatization movements provide a useful lens for understanding the synthesis of global and local frames because they are simultaneously shaped by global processes, including multinational corporate power and international financial institutions, as well as being driven by local understandings of environmental resources and local political context. On the one hand, they are influenced by the spread of globalization, including the push for neoliberal reforms at the municipal level, the pressure from multinational corporations to outsource water services, and the impact of international trade and investment treaties. On the other hand, they are intensely local, drawing on local constituents for mobilization, and dependent on the dynamics and structures of municipal governments.

Globalization and the Political Process: The Political Context of Contention

While globalization has reshaped social movement identities and frames, it has also transformed the political process, changing the relationship between movements and the state, creating new targets and a multi-scalar context of contention. These changes in the global political economy necessitate a rethinking of the current political process model of mobilization that has been the key theoretical approach for explaining contentious politics. The political process approach emphasizes the importance of the political institutional context—including the degree of openness of the political system, the stability of political institutions, divisions within the political elite and their willingness to accept movement claims, and the degree of repression by the state—for shaping the strategies and tactics of movement activists and providing the opportunities that influence their success or failure. The state shapes both conflict—by providing grievances around which movements organize—and the alliances necessary for the mobilization to emerge and develop.[48] Movements operate within the political arena with the goal of influencing policy and achieving political representation.[49] Changes in political opportunities or constraints link collective action to the state, spur new episodes of collective action, and influence the way social movement organizations and actors mobilize resources, as well as their choice of tactics and frames.[50]

Most recent research on contentious politics and globalization utilizes the "political process model"—which emerged to explain social movements at the national level—to explain how opportunities and threats at the global institutional level have shaped the trajectories of transnational movements and the shift in contention from the domestic to the international realm.[51] The value of the political process approach is clear: It provides insights into the role of political context in shaping opportunities for mobilization and outcomes, and it offers a useful framework for comparing movements cross-nationally and for understanding how structural differences between countries shape the divergent paths of episodes of collective behavior. Cross-national comparisons of social movements demonstrate the enormous importance of the state for movement outcomes and suggest that movements whose goals align with the central interests of the state are more successful in influencing policy than those whose goals conflict with those of the state.[52]

Yet, there are limitations to the political process model, particularly for understanding localized mobilization in the context of neoliberal globalization and the regulation of environmental resources. Some scholars have criticized this approach for privileging the nation-state over other institutional scales and for favoring political structures over broader political culture or the agency of movement actors.[53] By placing the state at the center of analysis, political process theory is unable to adequately investigate the role of multilevel opportunities, including the role of global opportunity structures, such as international regulatory institutions and transnational movement organizations, and the interplay between domestic and transnational opportunities, or structures that lie outside of the polity, including economic institutions such as corporations.[54]

Understanding the Political Context Above and Below the State

The nation-state alone cannot provide a full accounting of the causal mechanisms that explain social change. Institutions and agencies external to the nation-state are critical for understanding mobilization in an era of globalization. Multinational corporations and international financial and regulatory institutions have emerged as both sites of new power and targets of resistance from below. Many social movement scholars point to the power of globalization to produce significant cultural changes—creating new global social relationships that increasingly shape social movements through transnational flows of ideas, resources, and networks. Global processes are changing the nature of protest in terms of organizational structure, cultural repertoires, collective identity, social networks, and the relationship of movements to the state. Social movements respond to these transnational forces with new ways of organizing in order to transform existing political structures and institutions.[55]

Jackie Smith's research on the 1999 protests in Seattle against the World Trade Organization (WTO), for example, examines the global protest movement from the standpoint of the actors and networks involved. Her research demonstrates that global forces have created new transnational movement actors whose targets lie beyond the domestic level. Activists targeted international institutions and used new mobilization frames that reflected critiques of the global trading system, including the lack of democratic accountability of international trade institutions such as the WTO.[56] Activists involved in these kinds of global protests reach beyond the boundaries of domestic politics to respond to global opportunities for collective action.[57]

The increasing power of international institutions suggests that social movement scholars need to examine the multi-scalar context of contention, including the powerful global political and economic system in which states and movements are increasingly embedded. Existing social movement theories that center on the state are limited in their ability to explain contention against neoliberal globalization, including how transnational processes shape domestic movements.[58] Research on globalization and protest provides evidence that multiple-level analyses are necessary for understanding and explaining how global processes play out across different social contexts.

Most research on globalization and protest examines the shift in contention from local or national to the international arena. For example, Sidney Tarrow and Doug McAdam refer to this change as a "scale shift" and investigate why and how mechanisms of contention are transformed into activism at the transnational level. They maintain that transnational activism does not automatically form out of either a new global consciousness or in response to internationalization of the political economy. Rather, they claim, it emerges out of local forms of mobilization through coalitions, frame bridging, and identity formation. Tarrow and McAdam argue that it is important to examine the mechanisms that underlie the process of scale shift to understand why some movements successfully spread upward from the local to the national or international arena, while others do not. They contend that scale shift is stronger and more sustainable with the presence of brokerage, which links together previously disconnected movements and builds new identities and forms of contention. As a result, protest moves beyond narrow geographic, organizational, or cultural boundaries to new scales of action.[59]

While Tarrow and McAdam's conceptualization of scale shift adds rigor to the understanding of how local forms of contention spread upward and shape activism at the national or international levels, what remains understudied is the reverse process: the downward scale shift from the global to the local arena. Studying local movements that draw on transnational forms of contention is important for understanding movements that are rooted locally, but implicated in global flows of institutional and corporate power. Examining the processes that underlie this type of mobilization also explains why some movements are successful in linking global issues with local concerns, while others are not.

The Local-Global Nexus
Social movements emerge both globally and locally in response to international processes, and this alters the way they are structured and how they develop. Looking beyond the nation-state as a unit of analysis illuminates how global forces play out in local contexts, including how new representations of power at the local level are responding to and resisting

economic globalization. Global forces are challenged not only through global-level struggles but also by the historical and cultural discourses of particular contexts.[60] Focusing exclusively on the nation-state ignores power from below—the autonomous forms of decentered or local power—and limits understanding of how it forms and operates. Researchers miss this double opportunity when focusing on either movements that experience a shift in contention from the local to the international level or those that operate only at the global scale. Global processes—such as the global economic and political reorganization emerging from the shift to neoliberalism—generate new opportunities and institutional targets and facilitate the formation of new networks of social movement actors to contest local problems and threats.[61]

The rise of global institutions, multinational corporations, and transnational organizations has coincided with the emergence of powerful subnational regions in the global political economic order. As Saskia Sassen argues, global cities—as centers of technology and innovation that attract an educated and highly skilled workforce—have become new sites of power, that often possess more leverage in a globalized economy than state or national level governments.[62] Beyond institutional power, cities have also become sites of powerful resistance movements, as activists seek to protect the local culture, economy, and ecology from the threats of a global capitalist system.[63]

Research on localist movements, for example, reveals that the *local* does not exist independently from other political or economic scales, but rather is embedded in a complex web of networks and institutions, at regional, national, and international levels.[64] Local movements are increasingly paying attention to the implications of global processes on local spaces and mobilize to protect their quality of life and the place they live.[65] Melanie Dupuis and David Goodman refer to this as "reflexive localism," in which activists alternate between targeting institutions at different levels, including at the local, state, federal, and global level. They argue that movements which recognize when to focus on the global and when to direct their attention to the local level are well positioned to create "an effective social movement of resistance against globalism."[66]

Scholars are increasingly recognizing the interconnections between local and global issues. In their study of local environmental activism in the United States, Kenneth Gould, Alan Schaiberg, and Adam Weinberg point to the link between local environmental problems and the global macroeconomic system. They argue that communities are increasingly engaged in a "race to the bottom"—including lowering standards for environmental protection and social programs and cutting taxes—as they are pressured to compete with other cities around the world to attract investment from transnational corporations. To be effective, they contend, local movements must mobilize at multiple levels, linking local issues with organizations and campaigns at the global level, and targeting powerful global economic institutions that are seen as a threat to local communities.[67]

In the case of water privatization, the rise of neoliberal globalization has heightened people's fears about the commodification of the commons and the ensuing risks to and loss of control over a vital and life-sustaining resource, spurring local trajectories of resistance.[68] The shift in power from local communities to global institutions means that struggles over water resources at the local level must take into account the global nature of the problem and draw on resources and institutional opportunities that lie beyond the local. At the same time, water is intricately connected to local geography and watersheds and thus local political institutional context is also critical to water governance. In order to understand the dynamics of local struggles against water privatization, we need to take into account this interplay between local and global processes.

As anti–water privatization movements are shaped by both global and local forces and represent a form of grassroots resistance to neoliberal globalization, it is unclear whether the traditional political process approach, with its emphasis on the state as the central target of contention, is adequate for explaining the complex interplay between global forces and local contention. While political institutions at the state level clearly play a definitive role in shaping these kinds of movements, through economic and regulatory policies, and thus remain a key target for movement activists, there is a wider context—external to the domestic polity—that is also important for explaining the emergence, trajectories, and outcomes of local social movements.

Institutions and structures external to the state, such as global economic institutions and flows of capital, are important for influencing state- and local-level policies and for shaping our lived experiences. To understand how the social world is shaped by multiple sites of social reality—at the local, national, and international level—it is critical to expand the unit of observation beyond the nation-state to examine how power is constituted at different scales of opportunity structure and to explain differences between local and international opportunities and their influence on social movements. Decentering the state as the primary focus of attention, as Warren Magnusson argues, allows for an examination of the multiple and complex systems of power by which societies are governed, including new urban and global politics.[69]

Domestic political opportunities are constrained by the global institutions and processes in which they are embedded. Global flows of capital, multinational corporate power, and international trade and financial institutions have increasing influence over domestic governments—national, regional, and municipal—and their ability to regulate and control local resources. These transnational constraints can also alter the outcomes of local movements and the way they respond to opportunities, including the shift in power from domestic political institutions to non-state institutions such as multinational corporations and international financial bodies.[70]

David Meyer argues that domestic political institutions are nested within larger international political context, which affects their ability to negotiate policies. The capacity of domestic political institutions to overcome the constraining effects of exogenous institutions and structures and operate autonomously, as well as the capacity for movements to contest policies locally, depends on the degree of "institutional slack" available. When decision-making power is transferred from domestic to international structures, the degree of slack available to domestic institutions is diminished and thus autonomy is reduced.[71]

Movements that understand this dual nature of opportunities and targets and integrate both global and local processes into their tactical repertoires are more likely to be successful in achieving their goals. By synthesizing local and global opportunities, movements are able to draw attention to the vulnerability of local governments in the face of global power structures and create a sense of local solidarity that brings together activists and elites under a common fate. By broadening the political process perspective to take into account differences between local, national, and international processes as well as the interplay between them, researchers can investigate the dynamics of local movements embedded in global processes, including how new opportunities, threats, and targets from the international realm affect mobilization and outcomes of social protest on the ground. Assumptions about the static, one-way nature of social power are being challenged, particularly at the local level. Local actors, when compared to activists involved in national or international movements, have a greater capacity to shape political opportunities and structures. Where there is less distance between political elites and social movements, power is less centralized and more fluid.

The Economy: Non-state Targets and Opportunities

The central assumption of political process theories—that all social movements consider state structures as the principal target for collective action and for providing opportunities for successful outcomes—means that other salient opportunity structures and targets are either ignored or minimized. In an era of economic globalization, many scholars argue that new targets beyond the state, including economic institutions such as corporations or global financial regulatory bodies and international trade agreements, are increasingly becoming key targets of protest and opportunity. As nation-states and other levels of government focus more and more on economic growth and restructuring, and thus are implicated in the global hegemonic market discourse, it is not surprising that movements are beginning to target economic institutions as well as the state when making their claims and demanding policy and regulatory change.[72] Transnational economic forces transcend the nation-state through the movement of capital and jobs, as well as by shifting the enforcement and regulatory power, including labor and environmental protection, from domestic institutions to international regulatory bodies and multinational corporations. The result is new problems and conflicts as well as new targets for mobilization.[73]

David Pellow's research on transnational environmental justice movements demonstrates that these movements increasingly target multinational corporations as well as political structures in their efforts to reduce the flow of toxic waste to marginalized communities. The political process model, he argues, fails to consider the broader role of political economy in shaping mobilization, which is particularly important in an era of globalization, when corporations have increasing influence over the state. He offers an extension to the political process model—the "political economic process perspective"—which "acknowledges the intimate associations between formal political institutions (e.g., states and legislative bodies) and economic institutions (e.g., large corporations and banks) and their engagements with social movements. The political economic opportunity structure stresses the extensive influence of capital over nation-state policymaking, regulation, and politics and views corporations as equally likely to be the targets of social movement campaigns."[74] This approach is useful for studying social movements responding to economic globalization, as institutions above the state level, including multinational corporations, increasingly become the target of counter-globalization movements.

While scholars are correct in identifying the relationship between economic and political opportunity structures, most of the research in this area focuses on transnational social movement organizations and campaigns, shifting the attention from the nation-state to the transnational sphere. Yet with the increasing power of international trade agreements on domestic policy making, the local context of contention also matters. At the local level, governments often have more power in the face of international trade and investment treaties because they are not the principal signatories of these treaties and because activists have more access to local policy makers. This means activists can protect local resources from the encroachment of international trade and investment through their capacity to influence policies and regulation by participating in regulatory and decision-making bodies and by their close access to political authorities.[75]

As a result of the complex interplay between transnational and domestic opportunity structures and the increasing importance of economic structures as both sites of policy making and as targets for mobilization, social movement scholars can no longer understand local environmental politics by only investigating processes at the local level, nor can the effects of transnational corporate and economic institutions be understood by only examining their effect at the global level. To understand local movements resisting neoliberal globalization, it is critical to look at the intersections of global flows with local spaces.[76]

Globalization and the Network Society: Relational Processes in Social Movements

For problems to be understood as collective rather than as individual challenges, social movements need to mobilize into networks of individuals and organizations. Research on the relational dynamics of social movements demonstrates that movements are made up of complex network structures that mediate between individual actors, organizations, and the polity, and that these shape collective identity and movement frames, increase mobilization potential, and facilitate coalition building.[77] Research on social movement networks also points to the importance of direct ties between movement organizations—both domestic and transnational—for sharing information, bridging frames, mobilizing resources, and influencing the political arena.[78]

Movements are not necessarily separate or discrete entities, but are often connected to one another through the existence of bridging organizations or individuals, with multiple affiliations across groups, who link diverse movement organizations and enable broad coalitions. During episodes of contention, bridge builders facilitate the linking of previously disconnected individuals, organizations, or movements—particularly those with prior relationships that were strained—using social ties and shared understandings of a particular grievance or problem, through a process called brokerage.[79]

Recent research on brokerage has demonstrated the importance of social bonds for building cohesion between individuals and organizations and facilitating cross-movement networks. For example, in their study of environmental movement networks that mobilized in response to NAFTA, Rhonda Evans and Tamara Kay find that brokerage operates across several levels: by linking networks between organizations and movements; by facilitating resource interdependence, through frame concordance; and through the use of shared tactics.[80] Similarly, in her research on Brazilian youth activists, Ann Mische underscores the importance of movements achieving the right balance between strong and informal ties to facilitate a cohesive and integrated civil society.[81]

What are the processes that explain how coalitions emerge and develop and how they influence movement outcomes? Why do similar movements take different trajectories in terms of network structures and what factors contribute to the creation of cross-movement coalitions? Focusing on the underlying causal processes and mechanisms that shape these relational bonds provides the key to understanding these differences. Beyond the presence of structural conditions—such as formal or informal social ties—the emergence of networks or coalitions within and between movements is shaped by cultural and social processes, including class, race, gender, or identity. Individuals and organizations make decisions about their interests and goals based on existing cultural codes and social positions that can either facilitate or hinder alliance building. *Bridge builders*—individuals who link previously disconnected entities—are critical for helping individuals and organizations reconcile social or cultural differences and find ways to work together cohesively.[82]

For example, Fred Rose's study of cross-movement coalitions between the labor, environmental, and peace movements reveals that bridge builders are critically important for facilitating inter-class movements and coalitions that link activists and organizations from the diverse movements. Activists—including those from the New Left movements of the 1960s and older radicals involved in New Deal politics—were able to identify shared understandings of a problem that allowed activists from labor, peace, and environmental movements to overcome the class-based cultural gap and work together toward a common goal.[83] Building on Rose's research, Brian Obach investigates coalition building between labor unions and environmental movements in the United States and the factors that facilitate their success. In his book, *Labor and the Environmental Movement,* he argues that, beyond the culture gap, divisions between the labor and environmental movements are attributable to differences in their organizational position within broader political and social structures. Labor unions tend to engage more often in electoral politics, while environmental organizations—partly due to legal restrictions—are more likely to eschew partisan politics for a more nonpartisan issue-based stance. Nevertheless, Obach argues that labor and environmental movements can and should find common ground, and points to the critical importance of coalition brokers—actors with interests in both movement sectors—for uniting the concerns of environmentalists and workers.[84]

While these studies make an important contribution to our understanding of the role of culture in shaping movement coalitions, most research on bridge builders focuses on individual actors who, through ideological affinity, broker relationships between organizations in different movements. Yet organizations can also facilitate bonds between individuals, groups, and movements. Beyond individual cultural codes, organizational culture also matters for either facilitating or constraining brokerage. My findings on anti–water privatization movements highlight the need to pay more attention to the nature of the bridge-building organizations involved in movement coalition building.

Movement coalitions shape policy outcomes by creating new political opportunities and favorable policy changes. Yet these policy outcomes of networks are under-examined in favor of a focus on the formation of alliances themselves.[85] Examining the political viability of movement coalitions would shed light on the role of broader social, political, and economic processes in shaping coalitions and their outcomes. Specific episodes of contention and the structural conditions that underlie organizational alliances have garnered much scholarly attention. There has been a lack of research on the broader coalitions involved in social and economic transformation; most research on coalitions and networks examines discrete movements focusing on single-issue campaigns, or in the case of research on community-labor alliances on workplace-centered campaigns.[86] At the same time, there has been considerable focus on how community-labor alliances facilitate labor revitalization[87]; yet the processes that determine how and why coalitions emerge, and how they inform broader economic and social policy decisions, remain under-theorized.

With the growing power of neoliberal globalization to reshape domestic social and economic policies and environmental regulation, some scholars argue that social movement actors need to build strong cross-movement (and cross-border) coalitions that unite previously disconnected sectors, such as environmental, labor, and social justice movements. These coalitions are necessary for building a strong countermovement and offering concrete and viable policy alternatives to the damaging social, economic, and environmental consequences of economic globalization.[88]

Coalition Building in Response to Neoliberal Globalization

While globalization has transformed the targets and frames used by activists, it has also sparked the rise of global networks of social movement organizations. The growth of transnational advocacy networks is shaped by the dematerialization of the economy and the growth of a service based information economy—linking together people and workers in a global economic system—as well as international flows of people and information that create new and multiple shared sets of ties.[89] The globalization of environmental harm also connects people together, within and across borders, through the sharing of ecological risk. Ulrich Beck argues that sharing global risks in the world risk society creates a "powerful basis for community" that is both territorial and non-territorial.[90] The formation of new identities through shared risks unites individuals and organizations across borders and becomes the basis for collective action.[91]

In the new information age, transnational networks of global justice activists have emerged to form a powerful counter-globalization movement. Global activists join together through networks facilitated by new information technologies, "where actions, images, discourses, and tactics flow from one continent to another via worldwide communication networks in real time."[92] Many scholars argue that this new global civil society has the potential to radically transform politics and create a new social order based on a democratic and networked society. United through communication channels and by a common cause, global activists move across geographic, social, and cultural boundaries, forging new identities based on the notion of a global civil society. These transnational activists connect the global to the local by linking transnational concerns with local grievances.[93]

Research on transnational activism highlights the success of global advocacy networks in shaping the outcomes of transnational struggles. Activists share information and construct frames to hold governments accountable and pressure policy makers to change their behavior.[94] For example, in his book *Dams and Development: Transnational Struggles for Water and Power*, political scientist Sanjeev Khagram examines mobilization against the Narmada River Valley Dam Projects. His ethnographic account of the struggle to block the construction of thousands of dams in central India reveals the growing power of transnational actors to influence policies and practices at the local level. Networks of global activists use internationally recognized norms concerning human rights and environmental protection to alter the practices of states, corporations, and international agencies and pressure them to integrate these norms into development policies. Khagram cautions, however, that success is largely dependent on the ability of global activists to include the voices and participation of local communities engaged in the struggle. He argues that transnational coalitions are most effective at challenging the power of states and corporations if they play a facilitating role for organizations and activists on the ground.[95]

Although many scholars have investigated the emergence of transnational movements that unite movements and activists across borders to respond to global issues, globalization also facilitates *local* coalition building.[96] So far, little attention has been paid to the networks of actors and organizations that emerge out of local counter-globalization movements. At the local level, the growing power of neoliberal globalization has the potential to bring together previously disconnected movements, including labor and environmental movements. Unlike their transnational counterparts, coalitions resisting globalization at the local level have the potential to create powerful and lasting movements because they are rooted in territorial communities and demonstrate a shared sense of fate. Local-global connections through individual and movement networks, frame bridging, and resource interdependence shape the emergence, trajectories, and outcomes of local movements, particularly those responding to global processes. Local movement coalitions that form in response to globalization offer the opportunity to examine the processes and mechanisms that enable or constrain such coalitions and how critical network processes such as brokerage are reconstituted through globalization.

Research on social movement coalitions in the context of neoliberal globalization also helps clarify the difference between social movement networks and coalitions. Mario Diani and Ivano Bison argue that social movement networks are characterized by a strong collective identity, while coalitions tend to lack such identity bonds and instead emerge through instrumental need in response to specific campaigns.[97] Yet there is growing recognition that in response to the deleterious consequences of economic globalization, many social movement sectors are shifting their attention from narrow, identity- or issue-based politics to focus on broader social change, challenging traditional understandings of identity politics and social networks. The increasing importance of broader issues of social and economic justice has the potential to unite previously disconnected social movements under the common cause of countering neoliberal globalization and its consequences.[98] In the case of water privatization—in which the intersection between environmental, economic, and social justice is robust—previously disconnected movements, including environmental, social justice, and labor movements, bridge traditional divides and unite under the framework of creating a viable alternative to neoliberalism and economic globalization.

Coalition building in the context of a new common politics of resistance is especially critical to examine at the local level because of the capacity of globalization to create strong collective identities among groups and individuals who otherwise would remain disconnected. While the implementation of neoliberal policies at the local level in cities around the world has impacted urban areas and reshaped local politics, it has also generated contestation from the grassroots as organizations come together to offer alternative visions and policies.[99] Localized resistance movements that recognize the multi-scalar nature of environmental politics—and make connections between global and local processes—increase their potential of success by creating and a strong collective identity and local solidarity in the face of global threats.

Conclusion

While most studies of social movements tend to analyze frames, political opportunities, and networks as discrete processes, treated separately, they cannot account for the emergence, development, and outcomes of anti–water privatization movements. This is why, to understand the similarities and differences between the anti–water privatization movements in Stockton and Vancouver, I employ a dynamic theory of contention that encompasses multiple levels of analysis, including local and global processes, and how they combine to shape contention. The success of local counter-globalization movements is contingent upon their ability to draw on multilevel frames, opportunities, and networks and link global concerns with local grievances. Explaining the dynamics of local global connections through a rethinking of current models of contention is the goal of this book.

Each of the subsequent chapters interrogates factors at different levels of analysis, including ideology and frames, political opportunities, and networks and coalitions. None of these processes alone reveals the full story about why the anti–water privatization movements in Stockton and Vancouver took divergent forms and outcomes, despite responding to similar threats and forces. But examined together, what emerges is a dynamic explanation of the complex and multiple processes that shape counter-globalization movements at the local level.

3

The Meaning of Water: "The Commons" as a Socially Constructed Discourse

I remember it just really hit me one day when I was taking a shower. It was a hot day and I couldn't wait to get under the water. I felt the water flow over me and it was such a relief and then I started thinking about how such a simple thing, for so many people around the world, well it is impossible for them to have that sensation of clean water running over them. And then I thought, "What if I can't afford this? What if my kids can't afford this?"
—Sandra Lopez, Stockton, California

The anti–water privatization movements in Stockton and Vancouver were shaped by external forces, including the neoliberal push for private sector investment in water services. But that is only part of the story. Their emergence is also explained by locally situated meanings of water. Although the two movements diverged in terms of organizational and political dynamics, activists in both places revealed similar cognitive and emotional understandings of the public nature of water. Place-based and visceral meanings of nature are key to understanding mobilization around environmental resources, especially those that are essential for life. In this chapter, I investigate how meanings of water are intertwined with our situated experiences—including the historical, geographic, and cultural contexts in which we are each embedded—and discuss how activists' imagined worlds shape their political beliefs and decisions to participate in the movement against privatizing water in their communities.

In Stockton, anti-privatization activists' understandings of water were shaped by the historical, geographical, and political realities of drought, delta pollution, and pressures for water diversion. In Vancouver, understandings of water emerged out of an attachment to proximal watersheds in local mountains and a desire to protect the valued watersheds that provide drinking water to the region; perhaps some of the cleanest in the world. Constructed meanings of water in both Stockton and Vancouver reflect the interplay between the commons and commodity, and ultimately shaped the political standpoint of activists in both places, that water is part of the commons and should remain in the public realm.

The Meaning of Water

Water is essential for life. It permeates every aspect of our lives. Individually, we need water to drink, to bathe, to grow food; collectively, we need it for agriculture, energy extraction, industry, and manufacturing. Unlike other resources, water holds deep emotional, social, cultural, and spiritual significance because it is needed to survive. In an age of globalization, where geographic, cultural, and social boundaries are increasingly blurred and where people, capital, and toxins flow across territories, water remains intricately tied to people's sense of place and belonging. Understanding these deeply held notions of water is key to explaining why people would so passionately mobilize to protect water in their communities and fight to keep it part of the commons; one of the last public resources on the planet.

In Stockton and Vancouver, anti–water privatization activists spoke of their emotional and spiritual connection to water and described its significance to their participation in the anti–water privatization movement in their community. Many activists interviewed revealed that had the movement focused on any privatization issue other than water they would not have joined the cause as readily. The emotional and contextual significance of water for mobilizing people to join a social movement is strongly present in the narratives of the activists in Stockton and Vancouver, as are the psychological, historical, and geographical influences of their cognitive understandings of water.[1] For many, discussing their participation in the water movement provoked an emotional response, revealing a strong connection to place, a testament to the unique geographical and historical characteristics of the community where they live. Others described water as the source of life, a means of survival, evoking a religious or spiritual connection to water and describing it as a gift from God. A desire to protect their local resources and ways of life as well as the belief in the sacred nature of water as a human right and public trust compelled activists in both communities to mobilize to protect the commons.

Stockton: Protecting the Ecology of Place

When Sandra Lopez arrived at work one morning, she found her colleagues discussing the news that the mayor had recently visited the

Stockton water treatment plant and had announced plans to contract out water treatment and delivery services to a private company. Her first thought was that "this is dangerous for the water in the Delta. We can't let this happen." A forty-eight-year-old biologist, she works as a federal wildlife protection officer and is responsible for monitoring water issues in the region. She joined the anti–water privatization movement in Stockton because she believes that water is the most important environmental issue both locally and globally. Lopez traces her environmental and social justice activism back to her "Berkeley hippie days" in the 1960s, when she was involved in the campus riots and in fighting for "a better world." When she heard about the city council's plans to privatize the water treatment plant, she was furious and immediately felt compelled to speak out against water privatization. "Standing up for what I know is right is really important to me. And water is just too important an issue to stay silent about," she explained, adding, "If I am quiet then I agree with what you say. I got involved because I didn't want to sit back and just silently say, 'Okay, sure, fine.' I wanted to at least tell them they were wrong." Lopez believes that water is the most important issue for local politics because it is connected to everything, "it is life."

For as long as she can remember, Sandra Lopez has felt an emotional attachment to water and the natural world around her. For many people in Stockton, opposition to water privatization was directly linked to their identification as environmentalists, which shaped their understandings of the fragility of ecosystems and the importance of protecting the environment for future generations. Lopez told me she strongly identifies as an environmentalist and believes that protecting the environment, "especially air and water," should be the highest priority for governments, "even more than the economy." Her commitment to the environment led her to pursue a degree in biology and to work as a conservation officer.

Sandra Lopez connects her environmentalism to the ecology of place. Since childhood, she has felt passionate about protecting and conserving water, having grown up in El Paso, Texas, where water shortages are frequent. Similar to many other Stockton activists, she described becoming aware of the importance of the natural environment, the connectedness of ecosystems, and the importance of water for "everything" because of where she was raised. "I became aware of how important water is and how we need to protect ecosystems, because I was raised in El Paso, Texas," she explained. Lopez described her childhood, living near the oil refinery, and how she and her family worried constantly about water pollution in the Rio Grande River. She described the feeling of living "in the desert," where water scarcity was always an issue and "the prices of food and everything are astronomical." Because of her strong sense of identity as an environmentalist and her fears about water pollution and shortages, Lopez was motivated to join the anti–water privatization movement in Stockton: "I think it was water that made the movement so important to people. I mean all the people in the environmental movement in California will tell you when you do polling issues related to water quality and health of fish that people pull out of the delta and whether their kids can go swimming in the nearest water, all of that just ranks way up there." Like many others interviewed, she described her feelings about water as the reason for her involvement in the movement: "Because water is just too important to risk in the name of some company making a profit."

The understandings of water expressed by anti–water privatization activists in Stockton demonstrate the connection between people's beliefs about nature and the social worlds that they inhabit and which reflect their cultural and social understandings of the natural world. Meanings of water are embedded in individuals' social practices. The significance of water to people's daily lives was a consistent theme expressed by anti–water privatization activists in Stockton. As is apparent in the opening words of this chapter, Sandra Lopez was acutely aware of the importance of water in her daily life. She went on to explain what it meant to her personally to be involved in the anti–water privatization movement:

It was something that was very fundamental. You know, like Maslow's hierarchy of needs. Water is number one after breathing. And I just realized how precious it all is and how important it was to protect our water. Not just for me, but for my children and grandchildren someday. So they can take a shower at the end of a long, hot day and know that it is clean and safe . . . So many of us took water for granted. I did too until that hot shower.

In Stockton, activists shared similar stories revealing their deep lifelong emotional attachments to water, describing the importance of protecting water as critical to their involvement in the anti–water privatization movement.

Kelly Jones, a former naval officer who currently works as an environmental manager for the city of Stockton, became involved in the movement despite the risk of losing her job as a result of publicly opposing a decision by the city council. She said that she joined the movement because of her strong belief in protecting water. Jones, a forty-five-year-old married mother of three children, grew up on the East Coast of the United States, where she always worried about contamination in the water systems. She described to me being deeply affected as a young girl when she heard stories about polluted rivers catching fire and witnessed dead fish in the Great Lakes on a family trip to Expo '68 in Montreal, Canada. Her childhood experiences with water pollution led to a lifelong commitment to water protection. She believes that water should be the "number one issue" politically because "it is the source of life for all things living." Jones explained her willingness "to take certain risks":

Water is something that is really, really important to me. I moved around a lot as a child and when I got to the Central Valley, I realized how important water was. I thought to myself, my gosh, water is so necessary because of the fruit on the trees, and what we can grow here. So I think because there is such a connection between everything here and water, the issue is always on everyone's mind. Here, people are really worried about protecting water and about water being taken and diverted down south. I was worried about that for sure. And especially if a private company got their hands on the water.

Her description of the importance of water illustrates the link between situated experience and understandings of water, demonstrating that perceptions of risk and the construction of water as a social problem are deeply rooted in attachment to place. Anti–water privatization activists in Stockton relied on knowledge from their daily lives and experiences to provide evidence for why water was a critical resource to protect from the encroachment of capital. Interpretations of the environment and what constitutes environmental risk expressed by activists in Stockton challenge the authority and legitimacy of the state to make decisions regarding the management of natural resources.[2]

For activists like Kelly Jones, a strong identification with environmentalism and local water systems was critical for their willingness to speak out against proposed privatization plans. As she said, "We knew that water was too precious to risk. I mean this is something that every citizen needs to use every day of their lives!" This sense of the importance of water held true even in the face of the economically rational cost-benefit analyses presented by elected representatives serving on the city council. "I know more about the importance of water to the local ecosystem than the people who were driving the decision to sell off our water," she explained. Jones told me that the proponents of privatization, including the mayor and the business community, presented "heaps of so-called evidence" that the city would save money and increase efficiency, and "that we were going save money on capital improvements and increase efficiency and that this would be a boon for the city of Stockton." She added, "They continued to argue that because the private sector can do things less expensively, we were going to be able to operate more efficiently and generate more income and revenue for the city. I knew they were wrong."

Despite the official institutional information, Kelly Jones and other activists in Stockton articulated meanings of water that emerged from an experiential understanding, based on socially embedded and sensory notions of the physical world, rather than from rational, scientific knowledge.[3] Anti–water privatization activists articulated their concerns about local jobs and water quality and their perceptions of environmental risk. These perceptions were often based on emotional and observational responses rather than on concrete scientific evidence. Jones worked closely with the employees of the wastewater treatment plant in Stockton, but when she heard about the plans to privatize the plant, her concerns extended beyond the fate of the workers. She was also worried about the risks to water quality, and said that she feared that a private company would not invest in the proper maintenance of the plant's infrastructure because it would "cut into their profits." These claims were directly related to her observations from traveling through the developing world, where she witnessed people unable to afford clean drinking water, and where the link between water quality and illness was made clear. Jones worried that similar issues could occur in her own community if a private company gained control of the water system and was concerned that Americans do not grasp the fragility of water supplies. As she explained:

People here in the United States just do not understand what it takes to make water safe. How important that infrastructure is. In other parts of the world, 10,000 people die a day from dysentery from bad water. Here, people can just turn on the tap. They think, "Oh, no it would never happen here." People in the United States don't understand that half of the world spends half of their lives just hauling bad water into their house. Now they probably don't care because the majority of those people are women, but nonetheless people don't understand that just the privilege of being able to turn on a tap—for bad or good water—requires so many people behind the scenes looking into the quality of that water.

In Kelly Jones's case, her previous travel experiences, including witnessing people's daily struggles to obtain potable drinking water, shaped her concerns that privatization would limit access to clean drinking water for some people in Stockton. Underpinning the fear of environmental risk expressed by Stockton activists was a deep concern about the impact of private capital investment on the local community. Jones continued by describing her fear about rising costs of water under a privately run system:

I think people are so oblivious to this sort of unseen utility that they don't even see the parallels between $4.50 gas prices and the price of water. They just don't get it. They are not making the connection here in the United States, and I'm afraid that they are going to allow people who get the connection, which are the multinational companies, and when they turn around and figure out that they are paying more for a gallon of water than they are for a gallon of gas, than they are for a gallon of milk, it is going to be too late.

These concerns of increasing costs and lack of infrastructure investment to protect the local water supply influenced people's opposition to water privatization in Stockton. Fears, beliefs, emotions, and passions are among the diversity of responses to water privatization articulated by respondents in Stockton. These are the emotive and creative ways activists interpret political opportunities and construct grievances.

The emotional and spiritual understandings of water expressed by respondents in Stockton were often rooted in the importance of locality. Many anti–water privatization activists in Stockton expressed a strong attachment to "place," and this connection informed their understandings of water and fuelled their opposition to privatization. People frequently spoke of the importance of water to protecting what some described as "America's agricultural heartland," and many expressed a reverence for water as providing the source of life to the region. The city of Stockton lies on the San Joachim delta, one of the most ecologically sensitive water systems in California.[4] Severe water shortages and years of drought in the region led many people to view water as sacred, and in dire need of protection. Many activists argued that a profit-driven water treatment plant would threaten the water quality of the delta. Charles Barlow, a fifty-two-year-old lawyer and environmental activist, explained the importance of the local geographical context in shaping people's understandings of water:

Water-related issues are always a really strong interest to citizens in Stockton, and most people understand that they are right at the edge of the delta . . . I mean people made the connection between a private company running the city's huge wastewater treatment ponds, and the fact that the ponds are right out there at the edge of the delta and if anything goes wrong, the spill goes directly into the San Joaquin River. I mean people get that part of it. And then having a multinational corporation like OMI Thames coming in trying to cut corners and cut back maintenance and the whole thing. People got all that.

Similar to their Vancouver counterparts, activists in Stockton evoked a sense of proximity to the water system to describe their concerns about water privatization and the risks to water quality. Barlow felt that the anti-privatization movement in Stockton was stronger than other movements that have occurred in the area because it related to water. He credits the local importance of the delta and water-related issues to mobilizing people to join the movement. He described how he has spent fifteen years fighting to protect the delta from pollution and threats of diversion, and how his desire to safeguard the water system motivated him to oppose privatization and join the coalition as a member of the steering committee. Similar to Barlow, many respondents' primary concerns about water privatization were expressed in local terms. Perceptions of the increasing proximity of environmental threats were frequently discussed by respondents in Stockton.

Many framed the problem in terms of the impact privatization would have on their day-to-day lives and on the local environment. Brian Spencer, an insurance broker who devoted countless hours over several years fighting to prevent water privatization and then to overturn the private contract, emphasized the agricultural importance of the region. He explained, "I think up here water is very sensitive in nature and people are very cautious of water. We live in a delta. We rely on water to grow crops." Spencer recently moved to Stockton from southern California, with his wife and five children. He owns a boat and frequently enjoys outings along the river with friends and family and considers the delta to be "one of the most beautiful places [he has] seen." Before moving to Stockton, he told me that he had not previously been involved in social protest and had always believed that political action should be confined to the polling booth. Yet, when he heard about plans to privatize the water treatment plant, he felt he could not remain on the sidelines because he knew "in his heart" it was the wrong course of action for the city of Stockton. Spencer described the strong sense of protection of people in Stockton toward the water in the delta and how those concerns shaped people's opposition to privatization. "We are very protective of water here," he said. "We know we only have so much and so we need to protect it from profit-driven motives. . . . Once you hand it over to a corporation, there is no turning back. The delta as we know it could be lost forever, and the implications of that are frightening." Fear of losing control of a vital and precious resource drove him and others in Stockton to oppose the privatization of the water treatment plant.

The narratives of place described by activists in Stockton reflect a valued and cherished way of life that is increasingly perceived as being under threat. Historical attachments shape people's relationships to the landscape, and ecological change can create a sense of loss of previous ways of life.[5] Many activists interviewed referred to memories of times when the delta was less polluted and the air free of smog. The idealizations reflect the role of the imagined past in shaping their present-day connection to the surrounding physical environment, including water. James Vivian, a sixty-year-old carpenter, has lived in Stockton for over thirty years. He has devoted his life to social justice causes, referring to himself as a "committed spiritualist populist activist." He believes strongly in the connection between environmental, social, and spiritual well-being. He told me that he remembers a time when "the river was clean and you could breathe the air. And you could see all the way down the valley to the mountains," and lamented that that time "is long past." Vivian's connection to the local landscape undergirds his belief in the importance of local community. He described his attachment to place, when he said,

I am involved here in this community, and I'm never leaving this community. I intend to die here. This is my absolute home. We've put down our roots here. We love the sense of community, the beauty of the delta, the joy of working to build a better place. We are deeply involved in community work. We are utterly involved in populist, democratic movements here, and we are not moving. This is our place. Our children are all within a few hours of here. We have hundreds and hundreds of friends here.

His attachment to and love for his community motivated his desire to become involved in preventing water privatization and protecting the local environment.

James Vivian's response reflects the importance of rooted experience that characterizes the attachment to local landscape expressed by many activists in Stockton. These historical and contextual meanings of place informed the feelings of concern and anger that many respondents felt toward political elites and the decision to privatize the water treatment plant, and shaped their belief in the importance of protecting water as a public trust. Their desire to protect local water resources created a perception of uncertainty and risk connected to water privatization, and contributed to a strong desire to protect publicly controlled water services.[6]

"A Gift from God": Protecting the Commons as a Human Right and a Public Trust

Beyond concerns about protection of local ecosystems and ways of life, activists in Stockton evoked broader themes around water, including spiritual understandings and the importance of water as a public trust and a human right. The outsourcing of water services was seen as an affront to these sacred values. Possibly an even stronger connection than place-based attachments was made by many respondents who described water as the source of life, evoking a religious or metaphysical association. Judith Smith, a fifty-year-old author and filmmaker, described the "visceral" importance of spiritual connections to water as motivating people to participate in the anti–water privatization movement. "A lot of people in the movement saw it as a religious issue. That privatizing water was in direct opposition to God's will." Smith was told by one worker at the sewage treatment plant, who is a member of an evangelical church, "that his minister was opposed to the privatization because this is the water of baptism."

Sacred understandings of water emphasize the realities of nature that for many respondents lie outside of the rational or scientific realm. Many activists referred to the religious importance of water as being critical to their opposition to water privatization. Edward Lugert, a seventy-year-old retired department store employee, told me that he believes development is a threat to local ecosystems, and has been involved for many years in trying to protect farmland and wetlands from encroachment by developers. Protecting water—"a gift from God"—is fundamental to his opposition to development and his involvement in the movement to prevent water privatization. When asked why he joined the anti–water privatization movement in Stockton, he explained,

I've been married for forty-five years and have eight kids and have eighteen grandchildren and the good Lord has been good to me. . . . I want to protect my home as God protects us. And protecting water is a huge part of that. I see water as a gift from the good Lord and it is my responsibility to protect that gift. My grandchildren come out and play on my acreage, and run and ride their little electric car in the backyard, and play on the swing on the trees, and I know it sounds corny and maybe it even is, but I really like it. And so I need to make sure they are safe. That our water is safe.

Lugert's understanding of water as a spiritual gift of life challenges the marketization of nature. For many activists in Stockton, the physical world, including air, water, and land, is considered a sacred trust, to be shared among citizens and protected from the power and greed of corporations.

Others evoked the notion of water as a human right and spoke about the importance of maintaining water as a public trust. Ken Bernardo, a fifty-two-year-old union representative and former wastewater treatment manager for the city of Stockton, shares the belief that water should remain a publicly run service with many of his fellow anti–water privatization activists. His desire for water to remain a "public trust that should never be handed over to multinational corporations, [and] should never be for profit" extends beyond his commitment to unionized jobs and employment security for plant employees—adding, "and I am a big believer in the competitive nature of the private sector." He believes "that it is a citizen's right to have potable water and to have the wastewater treated. And I don't think you play games with that for profit. It is a public trust, a human right, and needs to stay that way." Despite a general support for the private sector, Bernardo believes that the unique nature of water as a public trust should exempt it from commodification.[7]

Many of the respondents interviewed echoed Ken Bernardo's commitment to keeping water as a public trust, such as Edward Lugert's statement that "water is a great example of a public trust." Lugert described his concerns: "Well I think everybody saw what it was. It was a water grab. It was a grab to control natural resources. . . . We all have a right to that water. We all need water, we can't live without water, our homes are useless without water, our farmland is worthless without water, and you've got these big groups wanting to control that." He feels that "water belongs to the community" and worries that without strong community activism and involvement by local citizens in political decision making, they will lose control of water resources. Stockton's residents recognize the ecological risks that stem from the commodification of resources, especially water, which they consider necessary for life.

The responses of the anti–water privatization activists in Stockton reveal the complexities of people's understandings and emotional perceptions around water. Rooted in specific cultural, historical, and geographical understandings, beliefs about water were expressed with emotion, passion, and a strong belief in the importance of protecting the commons. These socio-natural relationships not only shape individual worldviews, but also influence local water politics, as emotional attachments to water motivated activists to contest neoliberal ideology and the commodification of water. In Vancouver, activists were equally passionate about the importance of water and the need to protect it as part of the commons.

Vancouver: Local Understandings of Water

It was a warm June evening in Vancouver in 2001, when Sandra Gibson, a forty-eight-year-old office manager for the regional office of the Citizens Action League, led a delegation of activists and concerned citizens to a Metro Vancouver public consultation session to express their concerns about the proposed plan to privatize water services in the region. She was nervous. Because of her previous work as a union organizer, she had some experience speaking in front of large crowds, but this time she worried that she would be overcome by emotion when reaching the microphone because of her strong feelings about water. Gibson feels both an emotional and spiritual attachment to water. She spoke to me of the beauty of a rain drop, the wonder of a waterfall, and how as a child she loved to splash in puddles and watch her reflection in the mirrored surface of a pond. "Water was magical. Beautiful," she said, when describing her childhood conceptions of water.

When Sandra Gibson first heard about Metro Vancouver's proposed privatization plans, she was outraged and "felt like crying." Although she has been involved in activist work for as long as she can remember, she told me that the anti–water privatization movement was different from any other campaign because it was about water. She described the emotional importance of water and its role in facilitating her willingness to join the movement:

There is something about water. [It's] an issue that is close to everybody's heart. When you look at the tape [of the public consultation session], you will see that people got up and cried because they were so emotional about water, and so nervous to say anything but they said it anyway. . . . I remember one person standing up at that mike on the third meeting and all she could do was cry; she was so nervous. And all she said was, "It's my water. Leave it alone." And she started crying. It was amazing. That's what mattered. It wasn't all these great speeches. It wasn't. It was the power of water and what it means to people. That's what made me go out there and speak.

Her response reveals the role of individual emotional and visceral attachments to water in motivating respondents to join the movement. Similar to many of the anti–water privatization activists interviewed for this study, Gibson described her emotional connection and feelings around water as being fundamental to her participation in the movement. Unlike other issues, she felt it was something she could easily relate to because "it means so much in my daily life. It isn't out there somewhere, it is right here." She described water as something exciting and powerful; as something that made people less inhibited to fight against privatization, and drew them together because of shared understandings and concerns. She explained,

If it had been about something else, something other than water, I wouldn't have had the same reaction. They came because it was water. It was just so easy for people to relate to and grasp. I mean, come on, you can think about water in a hundred different ways. You play in it, you drink it, your kids like it, you have a bath in it. I mean all those kinds of things people relate to. So I think that because it was water, it was easier to deal with even than trees. And, when I think it got articulated that we can't survive without water, water is life, it began to hit people how important it is for us all. You know, here I am thinking about water all by myself and there are all these other people thinking about it. That was important for bringing us all together.

Her description of water demonstrates the complex and multidimensional understandings that many activists held, while pointing to the potential for bringing together actors from diverse social worlds under a shared understanding of water's importance in people's day-to-day lives.

Water encompasses social, economic, and environmental concerns that cut across divisions between individuals and movements, including traditional left-right political divides. It has the ability to transcend the environmental realm and mobilize people who do not consider themselves environmentalists. Sandra Gibson does not call herself an environmentalist because environmental concerns are not "at the top of [her] list." Yet protecting water is her main priority because it "is more than an environmental resource" and ties into issues of social justice, one of her most passionate interests. Water facilitates the coming together of diverse groups of people because of its many dimensions, including its social, environmental, economic, cultural, and spiritual importance.

As with Stockton, activists in Vancouver found water carried a particular significance among many and had a broad ability to mobilize a diverse group of people. Mike O'Brian, the national water campaigner for the Citizen's Action League, an organization focused on social and environmental justice, told me that he believed the movement came together quickly and with so many diverse supporters because it was focused on water, which he considers one of the "easiest issues to organize around," especially in relation to privatization: "I think because it was about water, it made it that much more compelling. You know, as we've seen in other public services, even health care, that's a harder sell—[with activists urging] 'don't bring the private sector in.' But on water it wasn't a hard sell."

A forty-year-old father of two children, Mike O'Brian lives and works in Ottawa, some 2,800 miles from Vancouver. He traveled across Canada in 2001 to help organize Vancouver's movement against water privatization. Over the past fifteen years, he has worked as a campaigner and community organizer on issues of social and environmental justice, and considers the anti–water privatization movement in Vancouver to be one of the most successful movements he has been involved with. For him it was inspirational. In his opinion, no other movement since has achieved a similar "level of grassroots support and energy." He credits, in part, the local significance of water, describing it as "the kind of issue that hits you on your dining room table. Do I trust the water that is coming out of my tap? Water is a really nuts-and-bolts issue. It just resonates and brings people together. People just aren't going to chance it when it comes to water." Like many of his fellow anti–water privatization activists, O'Brian expressed the sense of concern attached to water privatization because of the daily importance water holds in people's lives. The emotional attachments described by many respondents in Vancouver reveal their socially and culturally constructed relationships to the natural world. Their connection to water extends beyond viewing it as an ecological resource to be managed and regulated. The daily interactions that we all have with water are socially and culturally embedded—from bathing to quenching one's thirst or being caught in the rain without an umbrella—and these lived experiences shape the meanings we attach to what we consider a precious resource.

Mark Spencer, a fifty-five-year-old elected official, has been involved in regional politics for over twenty years, including having served previously on the Metro Vancouver Water Board. In all his years in local politics, he has never witnessed a more contentious issue than the struggle over water privatization. He described water as one of the most significant issues for constituents in the Vancouver region, suggesting that it was more important to people than the economy: "Why was the response so huge? Because it was about water. I think that the gut issue, the issue that people reacted emotionally and viscerally to, was someone else controlling our water. Water is seen as such a right by people. It is seen as the source of life. Clean water is part of their local heritage." He explained that water is as much a social and cultural issue as a political concern. "The reason this [issue] had such resonance with the public really relates as much to . . . people's relationship with a natural resource like water as it does the political reality of three Ps [public-private partnerships]. We have never been able to repeat that reaction with any other issue. That is the power of water." Mark Spencer joined the movement opposing privatization because of his passionate belief in protecting water for current and future generations, describing his wish for his six grandchildren to experience the beauty of the local natural environment as he did when growing up in British Columbia.

Locally embedded historical, geographic, and political contexts matter for people's understandings of water. In Vancouver, water is central to people's daily lived experience. From the frequent rainfall, to the surrounding rainforests, the particularities of place drive how many respondents in Vancouver view and understand water. "Water is a big deal for folks here. People here think about water all the time. People can see the watersheds every day. So they know where their water comes from, and that helps them understand the importance of protecting water"—that is how Sean Becker put it at the beginning of his interview with me. He is a thirty-nine-year-old community educator and executive director for a local NGO that raises awareness about issues of globalization, human rights, and sustainability. He went on to say, "I think it is also—and I am getting a little ethereal here—because it is such a rainy place we have this affinity for the water we have, and I think it is because so much of it is on us! [Laughs.] But I do think that is how it plays out on a local level somehow. It is different here than other places in BC and Canada, because of the watersheds." Becker devotes his time to community education about economic globalization, organizing public meetings and designing educational workshops for school children across the Lower Mainland. He told me that water is the issue for which people most readily make the connection between privatization and environmental risk, particularly in the Lower Mainland, because of "the beauty and closeness of the watersheds." For many Vancouver anti-privatization activists, local watershed protection was clearly a motivating factor for joining the movement and for the politicization of water.

Those watersheds kept coming up. How many people in urban centers in North America even know what a watershed is? Yet, for many of the activists in Vancouver, the visibility and proximity of the watersheds was driving local water politics. "Water in Vancouver is quite political, I think, because you can see the watersheds, they are right there. And if you log in the watersheds, your water is dirty. There is a direct connection," explained Sherry Carruthers, a twenty-eight-year-old youth activist and environmental consultant for the regional government. She continued, "Here, we are very much like, 'Our water comes from there and if you do something bad to that then . . . It is all very direct. It is not like some aquifer somewhere or like in Toronto, where it is very distant, it's not visible." Carruthers moved to Vancouver from Ontario, Canada, when she was a teenager to join the environmental movement and add her voice to the struggle to protect wilderness in British Columbia. In the cities in eastern Canada where she grew up, suburban sprawl is rampant, and nature is a "long car-ride away." She credits the closeness of the natural environment in Vancouver for transforming her environmental ethic and shaping her desire to fight for the protection of the watersheds. While she described the geographic and ecological importance of watersheds in shaping the politics of water in the region, she also spoke of the historical significance of water and watershed protection:

We have the benefit of decades of history of people fighting to protect water and save the watersheds. People did a lot of work over many, many years stopping the logging in the watersheds, way before this privatization thing ever came up. . . . And so they had kind of politicized people. In the spring when the melt would happen they would just be like, "Well, the water is dirty, there are no trees to hold up the dirt." And so people were very aware and they knew that we needed to protect water and stop the watersheds from being logged.

Previous movements to protect local watersheds by stopping logging practices had fuelled many activists' desire to protect the water supply. In Vancouver, watershed protection informs many people's commitments to water conservation and to keeping water in the public realm, setting the stage for opposition to water privatization.[8]

A Divine Resource: Globalization, Human Rights, and Commons
The socio-natural relations that are constituted through local settings are critical to the formulation of our worldviews, including the understanding of water as both a public trust and a human right and to the contestation of particular systems of belief. People's attachments to water in Vancouver reflect a profound mistrust of the private sector and of global capitalism, as well as a strong support for public control of resources. Mark Spencer described how anti–water privatization activists in Vancouver felt: "It was a line they drew, and it was a line they drew where they'd been pushed into the corner far enough. . . . It was sort of like, 'Well we've got air and we've got water and nothing else. It's all we've got left. We're not giving those up.' So there was a certain degree of 'last stand-itis' that was attached to it." People needed to stand up for public water systems as one of the last remaining publicly controlled resources in the province. Spencer went on to explain that people in Vancouver who opposed privatization did so because they feared that private sector involvement would threaten water quality and supply:[9]

We recognize how blessed we are in Canada to be able to have this public water. People saw it as being a heritage item for Canadians and commodifying it was contrary to their values. . . . I also think that there were the kind of issues about water quality issues, that our water supply would not be taken care of properly and there is probably a belief on the part of the public that . . . the private sector is more likely to be negligent than a public entity. And in looking after something like water, if somebody is motivated by profit there is more chance that they are going to forget to do their duty, so there was that fear too.

Mark Spencer believes in the human right to water.[10] He believes that threats to water are not only ecological, but also sociopolitical, and explained his fear that future wars will erupt over water because "it is the source of everything." At the same time, he believes that those fears fuel the kind of passion and emotion that is necessary for political change: "So you need emotional issues because they are the kind of issues that get people's attention and make them open their eyes and get involved in their communities. And then they become open to the other issues about economics and issues about local control and flexibility and those kinds of things. So emotion is needed. It is needed to create change."[11]

In Vancouver, social-psychological and context-driven understandings of water shaped a sense of collective belief in the importance of water as a public trust and a human right. Many activists recognized the incompatibility of water protection and the neoliberalization of nature. They described water as belonging to them, as part of their heritage and part of a commons that should be protected from market forces. The rejection of the commodification of the environment by anti–water privatization activists reflects an understanding of the interdependence of humans and nature.

Sandra Gibson described the sense of public ownership that she and others in Vancouver felt about water:

I mean my gut response without intellectualizing it a lot is simply because people identify water as something that belongs to them, that it is not something you pay for, that it is there, that it is God given. It is right. Nobody thinks about walking down to a river and having to pay for anything about it . . . So I think that is why so many people got involved. They were just beginning to realize that there are trade issues and corporate issues. They were beginning to realize that there are politics involved in water. I don't think people realized that before.

For anti–water privatization activists in Vancouver, public control signified something beyond local government management and regulatory oversight. The idea of common ownership is rooted in an understanding of water as a source of life that should not be owned or controlled by those who wish to make a profit. Sean Becker explained the connection between the idea of a collective resource and the rejection of global capital control of water:

People were really concerned about the loss of local control over a collectively owned resource. It wasn't just that people wanted the government to run the system. It was a genuine belief that water belonged to everyone. It was part of the commons. It is a human right. That was the idea at a fundamental level. And because of that, people don't trust the free market to control water at a core level. Even a number of folks who adhere strongly to the free market principle, when you really get down to it on a core level, there is a fundamental distrust of multinationals controlling our water.

Becker's response reflects an awareness of the neoliberal implications for nature that was expressed by many activists in Vancouver. By centering their understandings on the concept of the commons, activists in Vancouver challenged the individualism of privatization and profit and politicized the issue of water as a symbol of anti-corporate power and as part of the commons.[12]

Conclusion

Initially, it may be difficult to understand or explain why people would so passionately mobilize against the privatization of a water treatment plant, the kind of facility that most of us are either unaware of or do not consider in our day-to-day lives. Fundamental to this understanding are our emotional, visceral, and sacred understandings of water. Meanings of water are constructed through complex interactions between social relations, practices, and the physical environment.[13] The responses of activists in Stockton and Vancouver reveal that water is deeply imbued with symbolic meanings of power, justice, and spirituality; for many people interviewed, water holds the very power of life. In both places, anti–water privatization activists' political understandings of water as part of the commons—a public resource that should never be commodified—were tied to their situated experiences, reflecting the local importance of water. Proximity to water systems, community embeddedness, and cultural and historical values and practices shaped the range of perceptions around water—including its sacred and precious nature—in each place. These socially embedded constructions of water created an emotional potency around water that challenged traditional power structures and the hegemonic discourse around the commodification of water and globalization, fueling the political mobilization to oppose water privatization.

In Stockton, threats of water shortages, the importance of water to the local agro-economy, and the risks of polluting the delta drove the opposition to private ownership of water. In Vancouver, the proximity of the watershed and past struggles to protect it, and fears about loss of sovereignty over water systems fuelled the desire to keep water in the public realm. We need to look beyond the immediate structural context of mobilization to examine how political beliefs and actions are tied to the way we interpret and make sense of the world around us, as demonstrated by these emotional articulations involving the obvious presence of the delta in Stockton and the watersheds in Vancouver.

Social actors draw upon broader cultural toolkits to construct political discourse and guide social action.[14] Examining these ideological processes is particularly critical when dealing with movements that focus on environmental resources; understandings of nature are intertwined with historical, geographical, and sensory meanings of place.[15] Anti–water privatization activists in both Stockton and Vancouver expressed emotional and spiritual connections with water that were tied to historical and spatial context, and consistently made reference to the need to protect and conserve what they consider to be a precious and threatened publicly owned resource. Although mediated understandings of water were different across the two contexts—shaped by the specific situated experiences and milieu of each place—activists in both cities expressed similar beliefs in the life-sustaining and sacred nature of water resources, demonstrating the importance water carries for all of us, across widely divergent regions.

Meanings of nature and environmental risk are complex and varying and permeate our day-to-day lived experiences. The concern over the privatization of water, including the lack of faith in the accountability and trustworthiness of private corporations articulated by anti–water privatization activists reveals the contested nature of water—between the concept of the commons and commodification—and challenges the classic Weberian notion that individuals act rationally in response to bureaucratic and institutional informational processes.[16] Individuals are motivated by more than rational-based thinking; perceptions and understandings of nature are guided by noninstrumental social-psychological and contextual factors, including emotional attachments and historical meanings of place. By drawing on place-sensitive and emotional understandings of water, activists construct their own sense of environmental risk and political discourse around resources, challenging the power of global capitalism and the dominance of institutional rationality. Although many of the sentiments, fears, and narratives of risk expressed by activists in Stockton and Vancouver may appear to be exaggerated and are not necessarily grounded in accuracy or rational calculation, what they reveal is the social, cultural, and spiritual power of water.[17] What is clear from people's visceral reactions to water privatization is how the outsourcing of water services—what many see as commodification of the commons—conflicts with people's multiple and complex values around water.[18] The symbolism of survival inherent in the notion of water as the source of life sows the seeds for mobilization to protect the water commons.

The deep emotional responses to water described by respondents in Stockton and Vancouver reveal the importance of place-based attachments and the material conditions of ecology in the study of environmental and social justice movements. These lie beyond structural mechanisms such as political opportunities, networks, and resources, and beyond the noninstrumental perceptions of environmental risk and sub-political challenges to dominant power structures. Environmental policies and resource management programs need to reflect locally situated meanings and social practices. To explain conflict over resources we need to understand how, through cultural symbols, we organize our relationships to nature.

4

Constructing the Problem: Framing Strategies

What really pushed us over the edge was being shut out by the council. We were opposed to water privatization because of the arrogance of the mayor and his cronies saying, "We know what we are doing. You are just too stupid to understand. We are doing this, this is right, and we are smarter than you. You elected us, so why don't you just shut up." They always figure they can just do whatever they want to do. The city of Stockton didn't follow the rules. They just simply said, "This is ours and we will do whatever we jolly well feel like." I would say it is an old cow-town attitude of the leadership within the city and the county.

—Sandra Lopez, biologist and activist, Stockton

There were a lot of people familiar with the whole international scene on water. They knew what was happening in Bolivia, in South Africa, and they shared those stories. We met some of the people whose lives had been affected by the corporate schemes of multinational water companies. That was a huge galvanizing point. I remember at one of the meetings, I stood up and said, "You know, if you wanted to kill a lot of people really quickly, I can't think of a better way than pricing water." Just look around the world and see what has happened in other communities. There are just so many people who can't afford it. So many things went wrong in those places. It's terrible. So we shouldn't include water in the equation. Never water.

—Sandra Gibson, social justice activist, Vancouver

We are more likely to listen to and identify with a story that is well told—even when compared to another story with similar elements to which we can relate—when a story touches our own lives. Well-told stories carry power, and as illustrated in the previous chapter, there are many stories to tell when the subject is water. The narrative that anti–water privatization movements need to construct about the problem of local water privatization must be coherent and present a unified, well-articulated argument to political elites. In this chapter, I move beyond the broader examination of systems of belief to investigate the role of strategically constructed movement frames.

How are the emotive and visceral understandings of water constituted into well-defined and clearly articulated frames that act to present the problem and offer viable solutions? Despite many shared elements—similar concerns about water privatization, a sense of water as a sacred trust that should remain public—the anti–water privatization movements in Stockton and Vancouver relied on divergent framing strategies. In Stockton, the frame used by movement actors was neither clearly articulated nor consensual, and failed to create the necessary opportunities for their claims to be heard. The movement in Vancouver, on the other hand, constructed a clear argument that resonated with political elites and facilitated opportunities for movement claims to be considered. The reason for this divergence between the two movements lies in the adoption of global frames by Vancouver activists that created a sense of local solidarity in the face of transnational institutional threats.

How are global processes—the power of transnational capital and international trade agreements—articulated and resisted at the local level by anti–water privatization activists? Global frames are a key source of symbolic leverage for creating solidarity among social movement actors and political authorities at the local level. In Stockton, activists did not use global frames and focused instead on local political accountability. As a result, the movement was unable to shift the political discourse about the issue beyond local tensions between political elites and anti–water privatization activists. A key framing strategy for the success of the anti–water privatization movement in Vancouver was linking global issues with local concerns through *global connectors*, because that linkage helped create a sense of local solidarity standing firm against global corporate and trade risks. In Vancouver, anti–water privatization activists drew upon symbolic anti-globalization and anti-corporate politics to foster local unity and leverage local political institutions to support their demand for a publicly controlled water system.

The responses by activists in Stockton and Vancouver demonstrate the different ways social movement actors articulated local understandings of water privatization and begin to reveal the divergent movement-building conditions that shaped their diverse trajectories and outcomes.

Stockton: Let the People Vote!

For most of her seventy-four years, Dorothy Thomas has been involved in political activism, from campaigning for civil rights in the 1960s to marching for gender equality in the 1970s and more recently fighting to protect the environment. When she heard that the city of Stockton was planning on contracting out municipal water services to a private corporation, she prepared herself for yet another battle against a "conservative, right-wing government." She described the "lack of democratic accountability" of the local government as the main reason she became involved in the movement, along with her belief in the right of voters to decide who should control and regulate water resources and the provision of services. As one of the key leaders of the coalition that formed in response to proposed plans to outsource the city's water services, Thomas urged the anti–water privatization campaign to focus on raising awareness of procedural issues and problems at city council and call for a public vote on privatization contracts.

"We labeled it democratic injustice," she explained and described how the anti–water privatization coalition in Stockton urged people to rally around the push for a ballot initiative on the privatization of services. Thomas considers open dialogue to be the cornerstone of democracy: "I primarily got involved because it was a closed process. I really believe in democratic discussion. It needs to be a dialogue between both sides, and that was not happening." As past chair of the local chapter of the Association of Democratic Voters, a national organization focusing on safeguarding systems of government and on influencing policy through advocacy, she joined the movement in Stockton to "stand up for the democratic rights of the citizens of Stockton." She described the lack of openness on the part of the municipal government in Stockton as driving her participation in the movement to stop water privatization, which "was being railroaded into existence for us by government. That was the primary reason why I got involved. I saw what was happening politically through our illustrious ex-mayor and it was those things that grated on my nerves and irritated me." Thomas resented the treatment of anti–water privatization activists by the mayor and council, and argued that "we were not stupid or ignorant or as inexperienced as they alluded to [us being]." She believes that citizens should be included in political decision-making processes, particularly when it comes to issues that are "critical to life," such as air and water. In her opinion, the best way to allow people a voice is to provide them with an opportunity to vote on issues.

Many respondents in Stockton echoed Dorothy Thomas's concerns about the lack of transparency on the part of the local government. Activists' shared understandings of water as a communally owned resource shaped their claim that authorities should allow citizens the right to have a say in the delivery of local water services. John Sandler, a former maintenance engineer at the Stockton Municipal Utilities Department, who lost his job after the water treatment plant was privatized, argued that the biggest threat from privatization was the relinquishing of local democratic control of what he calls a "vital resource." He too was disillusioned with the lack of public consultation around water privatization and eagerly joined the anti–water privatization coalition to help collect signatures for a ballot initiative that would allow citizens to vote on privatization. He also helped organize demonstrations at city hall where activists targeted local politicians, shouting the slogan "Let the people vote!" According to Sandler, the problem with privatization was that the government would be handing over the responsibility for protecting resources to a private company. He explained, "Before, when there was a problem, a citizen could take it all the way to the city council, and they were accountable to the people. Now, if they have a concern, they are connected to a call center halfway around the world. How is that for accountability?" During the interview, Sandler articulated his strong belief in the public nature of water, noting that "the main arguments were that this is ours, it belongs to us, and we should have a say-so in what we are going to do with it"; and the need for voter approval on water service contracts: "If privatization is so good, present all the information to us and let us make the decision, through a democratic vote, not a panel of seven council members. It's a $600 million facility there and it belongs to us, and so we were against the fact that only a few people were leading the charge to sell off our utility."

Anti–water privatization activists in Stockton did not view the state as benign, and what they had to say about the democratic process demonstrates this fraught relationship between citizens and political institutions. Dave Alexis, a fifty-six-year-old political activist and former elected representative, identified discontent with government as a critical factor for movement mobilization in Stockton. "I think the main reason why people were so upset, including myself, didn't even get to the issue of privatization or why it was wrong. It was the lack of process that infuriated the public and created the backlash," he explained. A small-business owner and member of the local chamber of commerce, he has voted Republican all his life, and self-identifies as a conservative. Yet he opposed the decision by the conservative mayor and council to outsource water services, because he believes that water should remain in public control and that decisions about resources should require input from the citizenry.

Dave Alexis argued that the mayor and council "were totally derelict in their responsibility to the public. The public was not allowed to interact in any way. They simply refused to let people have a say, and I quote the former mayor, who said, 'This is a real sophisticated issue, and the citizens can't understand it.'" When Alexis realized that a majority of city council members wanted to fast track the vote on privatization, he was furious and felt that the municipal government was violating its responsibility to allow deliberation by the public. During a meeting he requested with the mayor, he asked him to delay the vote so that there could be further study of the benefits of outsourcing. The mayor refused. Alexis explained that because of the actions of the mayor and council, the issue of concern for the anti–water privatization movement became increasingly focused on the lack of democratic process, rather than on the negative consequences of privatization: "Framing-wise I think the fact that some elected officials just think people are stupid was a huge factor. It created the perfect storm."

Dave Alexis told me that municipal politics in Stockton had long been characterized by a division between conservative and liberal ideology. The historical context of political conflict fuelled the anti-privatization sentiment of the movement in Stockton because the decision was linked to a broader sense of discontent and mistrust of the local government. He described the "general sentiment of anti–city council, anti-politics in Stockton" and told me that past criticism and negative views of the mayor and council by key movement leaders had facilitated mobilization against privatization plans: "It wasn't only water privatization, there were other things going on with the development community, and so everything was way out of whack. There was the arena at the ballpark, which went way over budget. There was more waste during that tenure than we probably had seen in the preceding fifteen years to the magnitude of $50–$100 million in waste." Issues beyond water privatization and its consequences created an atmosphere of mistrust among movement actors and shaped their views of elected officials as corrupt and irresponsible.

Activists' understandings of the political process were instrumental in shaping key movement frames. For people in Stockton, risks of water privatization were discussed in connection with the divisive culture that characterized the political arena historically in Stockton. Individuals drew on existing political cultural codes, including traditional left-right schisms, to interpret the problem and construct arguments. James Vivian, author and social justice advocate, was deeply dissatisfied with the city council members at the time of the privatization battle. His unease extended beyond the issue of water privatization to encompass concerns about greed and political corruption, describing the city council at the time as "the most crooked council" he had ever witnessed. He explained that the movement's main arguments were related to fears about a lack of democratic process in local politics. "It became less an argument about privatization than one about democracy and high-handedness. It became about the abuse of the democratic process. What they did was legal, but devious and furiously undemocratic," he argued. Vivian described how the movement targeted political corruption rather than concerns about corporate control of resources.

Anti–water privatization activists linked their grievances to broader concerns with neoliberal ideology and the Republican legacy of the previous two decades that created deep divisions between liberals and conservatives in Stockton. Paul Conners is a forty-five-year-old urban planner who has lived in Stockton most of his life. He is an environmentalist and social justice activist who became involved in the anti–water privatization movement because he did not want to see local water services controlled by a private corporation. Yet his participation in the movement was connected not only to his concerns about corporate control of resources, but also to an "immense distrust of government in general that kind of lit a lot of the fires here." He described how a sense of political cynicism shaped an argument centered on local political corruption and accountability rather than on the corporate policies that threaten water.

Stockton has been a place where, for a lot of different demographic reasons, politicians have been kind of able to take advantage of the local citizenry. So we have a built-in skepticism and overt cynicism about anything related to government getting involved in issues. I think the reason people reacted so viscerally to oppose privatization is not just that they thought it was just a bad idea, that financially it was going to end up screwing the local people, it was also that they didn't trust the city council. Most people did not trust the mayor at the time, who was perceived as being another cookie-cutter Bush Republican type.

Movement actors in Stockton focused on privatization as a misstep by a corrupt municipal government, an argument that clearly delineated the opposition and weakened the strength of the movement.

Activists in Stockton located the problem of water privatization as grounded in local political processes that constrained participatory democracy. They rarely referenced the global nature of the problem or drew on arguments from transnational water movements, despite the fact that the companies short-listed for the contract were all multinational corporations. Most activists agreed that local political context, including dissatisfaction with the mayor and council and a feeling of skepticism around government competence, was the central focus of the movement. Few people in Stockton discussed global processes as being important to the movement's arguments. As Joan Davidson, a leader in the anti–water privatization coalition explained, "We did what we needed to do here. Some people say that we should be connected to something more global, but you know, sometimes it is hard enough to formulate a goal and to marshal the kind of interest locally; we need to reference what makes sense to people here."

Despite the fact that this view was articulated by the majority of activists in Stockton, the decision to focus on local political frames was not consensual among movement participants. The overlooking of global linkages caused divisions between movement actors in Stockton, with some people arguing that a focus on globalization would have increased movement mobilization and facilitated opportunities for grievances to be heard. "The opposition movement in Stockton was unconnected to anything global. But the movement in favor of it was connected to everything global. The financial investment was global, the companies competing were all foreign, and they had a huge influence over the mayor and the rest of the city councilors," explained Kyle Winters. A former city bureaucrat who resigned his position during the battle over privatization, Winters believed that linking the movement's argument to global concerns would have allowed for a strategic counter-framing plan. His experience as a political consultant led him to question the efficacy of the anti-government stance by activists in Stockton, which he felt alienated potential political allies who might otherwise have sided with the movement. He said that he felt that the anti-water privatization movement should have focused on the consequences of privatization on local control of water in light of the track record of the multinational water companies that were bidding for the local contract. During the interview, he described how the proponents of privatization, including the mayor and the business community, used global frames to convince the public that the outsourcing decision was in the best interests of the community. "They told us that global companies had the expertise and the financial capital to invest in our water system and make it more efficient and that they would save the city money and transfer risk from the taxpayer to a private company," he explained, arguing that anti–water privatization movement should have focused on global-level concerns to counter and neutralize these arguments.

While many activists argued that downplaying global concerns constrained their ability to neutralize claims by the municipal government about the expertise and economic efficiency of multinational water companies, others expressed their frustration that neglecting the global dimension negatively affected the movement's ability to mobilize. Jeremy Beck, a local writer and teacher, told me that he was disappointed with the lack of young people involved in the movement in Stockton. He felt that mobilizing youth would have been relatively easy because of the university population. Beck recognizes the effect of neoliberal globalization on local communities, and discussed his concerns about the outsourcing of jobs and the "race to the bottom" in terms of social and environmental policies that he sees as the consequences of globalization. He explained that had the movement tied its opposition to water privatization to larger critiques of global capitalism, they could have increased their mobilization potential, particularly among youth with ties to the anti-globalization movement:

I often wonder if things in Stockton would have been different had more young people been involved . . . I think it would have helped if they had made reference to the anti-globalization movement and the connections to the bigger picture. Instead, there was little that resonated with them [youth activists], little that would help mobilize this group of people. It could have been sexier if it had been connected to a massive global movement the way it was with Vietnam. Then you could do community organizing and make people feel part of a global revolution. That could effect change.

Jeremy Beck explained that there were several youth activists initially involved in the anti–water privatization movement, including one young woman who had recently attended the World Water Forum with a group of global water activists to protest the commodification of water by multinational water firms. He told me that these activists wanted to connect the local movement to broader anti-globalization movements and raise awareness of the global risks to local water privatization. According to Beck, these youth activists were discouraged from focusing on the global nature of the struggle by other members of the anti–water privatization movement and, as a result, their participation in the movement waned. By minimizing the importance of global issues, the movement prevented a broader range of actors from participating in the struggle against water privatization, including those involved in anti-globalization work.

The focus on localized political frames left questions of neoliberal globalization and the power of transnational capital to undermine local control of resources unexamined by respondents in Stockton and shifted the focus from the problem of privatization to the conflict between citizens and elected officials. Activists downplayed the ability of local governments to oversee public services such as water treatment and delivery, and demonstrated a high level of distrust in government, regarding it as part of the problem, rather than part of the solution. There was little faith among respondents in Stockton in the capacity of political institutions to regulate and deliver services effectively.

One of the reasons why the movement in Stockton located the problem in the antidemocratic nature of the municipal government is related to situated experiences of key movement leaders. Many of the activists involved were connected to organizations that focused on electoral politics and voter initiatives at the domestic level. In California, social movements often use direct democracy measures, including voter-driven ballot initiatives, as a strategy for policy change. Citizens groups use the initiative process to bypass traditional representative institutions, including city councils and the state legislature, to propose and vote on new legislation.[1] Many of the respondents in Stockton described being involved in voter initiatives in the past and discussed the importance of direct democracy to policy change. One of the main organizations involved in the anti–water privatization battle, the Association of Democratic Voters, devotes most of its time and energy to voter advocacy and legal challenges to political decisions. Strong ties to domestic organizations that regularly rely on the ballot initiative process focused the movement's attention on local political institutions and detracted from concerns about global threats. The organizational history and culture of key movement leaders also meant that there were few people on the ground who could act as bridge builders to transnational movements or frames.

Phyllis Kennedy, a community organizer with the Citizens Environmental Alliance (CEA), a national organization that focuses on the link between corporate power and environmental protection, described how she attempted to unite the anti–water privatization movement in Stockton around a common theme of global corporate power. She and other California-based campaigners with CEA visited Stockton when they heard about the proposed outsourcing of local water treatment and delivery services because they considered what was happening there as part of a strategy by multinational corporations to gain control of water systems in the United States. Kennedy is part of a global network of water activists and regularly attends international meetings related to global water issues. She explained that the organizers from CEA tried to shift the focus from local electoral politics to global institutions in the Stockton case in hopes of bringing to light the risks to local control of resources from multinational corporate power: "I don't even know if the local groups that got involved in Stockton even realized those global connections. You know, in our view, these were really nice people and very hard working, but they just weren't used to dealing with these big national or international or multinational corporations."

Activists from outside of Stockton expressed frustration at the unwillingness of local movement actors to embrace global framing strategies. Part of the reason for this resistance is that few of the people involved were connected to groups operating outside of the local context, such as the global movement for water rights or anti-globalization movements. Although many activists in Stockton were aware of the global issues, most downplayed their importance to local events, with some even expressing resentment at the CEA's attempt to alter their arguments. Dorothy Thomas said,

You know I really think the local coalition was the significant leader. The CEA came in and tried to help us. They were helpful in many ways, especially in helping us to become more of an organization. But them being outsiders, they were very unfamiliar with the situation in Stockton, not knowing what the climate was. They had some ideas that just didn't go over with the group, and so we kind of evolved from that, and we took the reins in our own hands.

Activists in Stockton resisted the influence of external actors and were reluctant to trust outsiders who encouraged them to incorporate global issues into their arguments. Their sense of localized agency was strong. They saw themselves as good, moral citizens who were capable of making decisions and guiding policy that would affect the community.

Relying on the overarching frame of antidemocracy, activists in Stockton articulated the problem as being rooted in local political institutions and the culture of the political process. As a result, the movement reinforced the division between citizens and elected officials, which prevented them from communicating a clear solution to the problem of dealing with infrastructure upgrades and proper maintenance of water resources and service delivery. The movement frames in Stockton excluded issues that had the potential to mobilize a wider pool of movement recruits. The anti-government frame also served to alienate the political representatives who held the decision-making power, pitting activists against state actors and preventing potential ties with political allies. As a result, the initial organized movement to block water privatization in Stockton failed.

Tom Bailey, a local businessman, friend of the former mayor, and one of the proponents of privatization, explained how the anti-government focus worked against the movement's interests:

I think they vilified the mayor, and basically drew a line in the sand, and from that point they couldn't retreat. I knew some of the people involved, and I told them to look at it from a pragmatic, private sector mentality, and my counsel to them was to quit vilifying because you are dealing with a popular mayor and the more you vilify him, the more he gets press time in the newspaper. And they didn't listen to me. It was an uphill battle for them. They might have a better movement to protect the utility if they had countered the mayor's arguments and not just gone after him on a personal level.

Respondents in Stockton translated their visceral concerns about water and their belief in public control into frames focused on voter rights and political accountability. The strategic choice made by the movement organizers to focus on the abrogation of democratic processes shifted attention away from global threats to local control of resources and alienated political elites, preventing the movement from neutralizing pro-privatization arguments and closing off potential avenues for public input on water privatization. Hence, anti–water privatization activists in Stockton failed to prevent the outsourcing of water services, and were forced to engage in a long, costly, and protracted legal battle to overturn water privatization.

Vancouver: Connecting the Global to the Local

While the movement in Stockton remained squarely centered on local political issues, the movement in Vancouver used global frames to create openings for claims to be considered, to foster local solidarity, and to mobilize. The use of global frames reflects the diverse movement-building conditions—including political and organizational culture—in Vancouver as compared to Stockton. For activists in Vancouver, concerns about the commodification of water and a belief in the importance of protecting the commons and safeguarding local democratic processes led them to frame their concerns about proposed water privatization around the risks to local control of water systems from multinational corporate policies and international trade agreements. Vancouver respondents frequently described the connection between the local push for outsourcing water services and the consequences of water privatization in communities in other parts of the world.

How do global issues become incorporated into local movement frames? While scholars have identified the connection between local and transnational frames, the specific mechanisms that underlie the global-local frame-bridging process remains unclear.[2] In Vancouver, the presence of key movement leaders with connections to and personal experience with transnational movements and institutions shaped the on-the-ground framing strategies of the campaign. In Vancouver, many of the activists involved in the anti–water privatization struggle had previously been involved with other movements whose focus was on global issues, including the 1999 anti-WTO protest in Seattle and the global campaign to stop the Multilateral Agreement on Investment (MAI). These individuals acted as *global connectors*, adapting preexisting institutional frames to new sites of collective action.

Sean Becker was one of the global connectors in the Vancouver movement. As executive director of a local nonprofit community organization that focuses on youth education around issues of global social and environmental justice, he was involved in the international anti-globalization movement, and traveled to Seattle in 1999 to participate in the anti-WTO protests. He has regularly attended international conferences and meetings on issues of globalization, trade, and social justice. Through his anti-globalization activism, Becker is linked to global networks of organizations and individuals. At the same time, he is connected horizontally to activists and organizations on the ground. His work educating and organizing youth around issues of globalization facilitated the diffusion of frames from the anti-globalization movement to the local anti–water privatization movement. "For me, it was a real, local, concrete example about the things that we were talking about at an international level," Becker said, explaining his motivation to join the movement. "We pointed out that what was happening here was a manifestation of what we had identified as the problem on a global scale. And we said, 'See, this is what it looks like.'" He added, "It was that immediacy that made it exciting for people to be involved because sometimes it can be a hard connection to make locally when we are talking about globalization and international issues."

Similar to Sean Becker, many leaders in the movement in Vancouver were embedded in global movements, both structurally through organizational and network associations and social-psychologically through preexisting anti-globalization frames and identities. These leaders also had close ties with local movement organizations and activists. Their dual global-local identities made them key leaders in the anti–water privatization movement because of their ability to draw on preexisting arguments to make sense of what was happening on the ground and use those frames to engage a broad-range of supporters.

Global connecters literally translated global issues down to the local level, adapting existing anti-globalization frames into concrete and easily understood arguments that resonated across social movement sectors. "It was a good experience for me also around really understanding the importance of the municipal level and the degree to which we can connect some of the global and national stuff to the municipal level. Because a lot of these things we talk about in terms of global trade agreements actually boil down to the municipal level," said Fiona Rogers, whose previous activist and educational experiences focused on the effects of globalization on communities in other parts of the world. "I have done a lot of different activism that has been more on global issues, in Latin America and other places, and I felt that the water privatization issue was the first time I really, really dedicated myself to a super-local issue." For activists like Rogers, engagement in the anti–water privatization movement stemmed directly from their involvement in global movements for social and environmental justice and their ability to recognize the local manifestations of global problems.

While the involvement of global connectors was critical for how the Vancouver movement constructed its arguments, framing processes were also shaped by other structural factors, including the organizational characteristics of the groups involved. Two of the key organizations involved in the anti–water privatization battle—the British Columbia Public Sector Employees Union (BCPSEU) and the Citizens Action League (CAL)—had organizational structures that linked them to national and international networks of activists working on issues of water privatization.

The connections with national and international organizations and resources shaped the contextual focus of the movement in Vancouver, bringing a global perspective to the local issue. "My international outreach work on water was really important in shaping the role I played in the local movement in Vancouver," explained Mike O'Brian, water campaigner for CAL, a national organization with local chapters in the Vancouver region. CAL is also part of an international network of water activists who regularly share resources and exchange information. These global water activists, including O'Brian, met regularly at international events, such as the World Water Forum, an international conference on global water issues held every three years. He told me that attending these meetings allowed him to gain a solid understanding of the connections between neoliberal globalization and the commodification of water. The organizational structure of CAL, with its integration into a global network as well as its strong presence at the community level, facilitated the diffusion of global frames to the local movement in Vancouver: "Being connected to a global movement really made me and others who were involved realize that there was a new pitch for water. That the same thing was happening all over the world in different communities, with the same players—the same companies making the same pitch." O'Brian explicitly described his role as a global connector: "We instantly realized that we had to make international connections, and we have to get money and resources and analysis to local groups." The organizational structure of

CAL was critical for shaping frame dynamics in Vancouver by providing key informational resources that connected the local movement to global processes and events.

Framing water in global terms simultaneously connected people from the Vancouver local community to people from around the world experiencing similar struggles and drew attention to the risks associated with privatization by providing narrative accounts of the negative consequences of water privatization. At community meetings and at the public meetings organized by the Metro Vancouver Board, members of CAL distributed and asked people to sign the "Cochabamba Declaration," a document created at an international symposium on water privatization in 2000 that took place in Cochabamba, Bolivia—the site of one of the most publicized cases of mass uprising, which occurred after the water system was privatized in 1999.[3] The declaration referred to water as a sacred resource and a human right and called for an international treaty to protect water from commodification.

Many activists in Vancouver spoke about the sense of solidarity they felt with other communities around the world that were engaged in similar struggles. They articulated frames about global struggles and demonstrated a strong awareness of global initiatives for keeping municipal water systems publicly controlled. These local-global connections were facilitated by the presence of a key activist from Bolivia, who led the anti–water privatization movement in Cochabamba, which took place between January 1999 and April 2000. The Citizens Action League flew in Oscar Olivera, one of the main leaders in the struggle against resource commodification in Bolivia, to speak to activists in Vancouver about the experiences of people in Cochabamba and the dangers of water privatization.

"Hearing Oscar speak was very moving," said Eric Robinson, labor organizer and water researcher. While he had never traveled to Bolivia or any other community outside of Canada where water privatization had occurred, the opportunity to hear the stories from Cochabamba firsthand, and the connection he felt to a global community, motivated him to become involved in the local anti–water privatization movement. He told me that global water privatization narratives were, in his opinion, critical to mobilizing people in Vancouver to join the movement because they fostered a sense of solidarity with international communities. "Oscar was such a great speaker. He said that people who had their own wells had to put meters on them so that they could pay for the water. From their own wells! Wow, it really had a huge impact on people here. You could just feel people getting worked up and ready to fight," he said, describing the powerful effect of Oscar Olivera's speech. Robinson explained that the connection to struggles in other communities around the world gave people a sense that they were involved in an epic campaign.

Organizations in Vancouver explicitly sought to connect the local movement to similar struggles internationally to motivate activists and mobilize a more diverse range of actors in the fight against water privatization. "We learned a great deal from other jurisdictions about the companies that were purported as being the proponents for our water system here," explained Peter Clark, an environmentalist, and director of City Green, a local environmental organization in Vancouver. He added, "Sharing stories with other communities allowed us to look and see what they were doing elsewhere and get the true story from local jurisdictions." Global narratives not only helped create a sense of solidarity for local activists with people from around the world experiencing similar pressures to privatize publicly owned utilities and resources, but also provided concrete evidence of negative consequences of water privatization that could be used as leverage by local movements.[4]

Peter Clark spent more than ten years working in Africa and witnessed firsthand the struggles of people who had no access to clean drinking water or could not afford to pay for critical resources. During the anti–water privatization movement, he worked with City Green and other community leaders to raise awareness of the impacts of privatization, organizing lectures, and speaking out at the public consultation sessions. He told me that he joined the fight against water privatization in part because he feels a strong connection with people in different parts of the world who are experiencing similar struggles. He explained that, unlike any other issue, water generates solidarity among global communities because "no matter where you live, it is easily understood as a basic necessity." Clark described how the strategy of bringing to light stories from other communities facilitated movement mobilization:

Because there are so many horror stories from other places, we thought, okay, this is a place where we can take this on and basically really jam the issue. And we had done quite a bit of research on the companies that were the potential bidders of it, and we brought up people from Bolivia to talk about what had happened in Cochabamba in terms of the privatization of their water. And that really helped us build the sense of a movement here and we really went for it.

The strategic use of global narratives strengthened the movement's ability to mobilize by demonstrating that the issue was part of a larger global struggle and creating a sense of international movement solidarity.[5]

Global frames target a wide range of constituents because they resonate with people who focus on social justice and anti-globalization issues. David Smith, economist and social justice advocate, explained how the anti-privatization movement in Vancouver was linked, through a sense of global solidarity, to the anti-globalization movement that had a large and active membership in the region. He said, "The connection with a global movement was definitely huge. There was a really strong anti-globalization movement in Vancouver—in BC—at the time, with the 'Battle in Seattle' and the Stop the MAI campaign. So those people really became mobilized in the water fight, especially when those connections were made with what was happening globally." For activists like Smith, the broader links between water privatization and economic globalization were critical for motivating them to participate in the movement at the local level.

David Smith and his wife were deeply involved in the anti-globalization movement in Vancouver throughout the 1990s. They had traveled extensively in Latin America, working with local groups to fight neoliberal policies being promoted by international financial institutions such as the World Bank and International Monetary Fund. He explained that "although they were mostly concerned about impacts here, [people] also had a lot of empathy and concern and solidarity with what was happening in Latin America or Asia and to a lesser extent Africa. . . . People were really focused on the fact there were large global corporations and trade agreements that were facilitating their ability to just move their goods and investments around often with detrimental impacts on the local populations." Beyond narratives, Smith's comments and his observation that "just as globalization has created alliances among corporations, you are starting to see a lot of social movement players around the world starting to talk together and glean more together," reveals the importance of local-global alliances of activists and organizations in response to the consolidation of global capital and transnational financial institutions.

In Vancouver, the frames utilized by activists situated the problem of local water privatization in the context of the broader political economy of advanced capitalist states, including the social and ecological inequalities produced by the constant need to increase production and consumption.[6] The use of anti-corporate frames, with a specific focus on the multinational corporate agenda and the threat of expanding neoliberalism, created greater solidarity among domestic and transnational social and environmental justice movements, linking the local community with individuals and movements in other parts of the world. Fiona Rogers, community organizer and educator, described how focusing attention to the track record of multinational corporations involved in water privatization shifted the contextualization of the issue to problems of neoliberal global capitalism, which helped galvanize the movement. "It really helped people understand that water privatization isn't just happening in Vancouver, but is something that it is part of a larger neoliberal ideological shift. The fact that they were short-listing the same companies that had created so many problems in Bolivia and South Africa and elsewhere—even places in the U.S. and France—was something that really resonated with the groups here and added fuel to the fire," she explained. The use of global frames inspired activists involved with the anti–water privatization campaign by increasing the importance and relevance of the local struggle in the broader context of neoliberal globalization and growing multinational corporate power.

The anti-corporate framing strategy in Vancouver reflects the growing resistance to transnational institutionalized power structures that is shaping local movements opposed to the shift toward neoliberal globalization.[7] Heather Harrison, researcher and organizer with the BCPSEU, has worked for many years as a community activist and educator, raising awareness of issues of corporate globalization. In recent years, she told me that she has noticed a greater awareness of international justice and solidarity among social movement activists in Vancouver. She described the growing understanding of the link between international corporate policies and social and environmental injustice and how it is shaping contentious politics:

I think that those corporate horror stories are really important for people who are concerned about . . . international justice and solidarity. But sharing those stories is also part of a longer-term project of creating global awareness. . . . I think there is a change, especially with the activists I have worked with and how they think about things. Now we consider the global context. I think there is a slow change happening in the world where I hope there is going to be increased awareness about the fact that our corporate policies have had a direct impact on people.

The responses from anti–water privatization activists in Vancouver demonstrate that global events and institutions, and the intersections between politics and the global economy, have growing influence over local movements and the construction of collective-action frames.[8] Global frames served not only to galvanize the movement in Vancouver, but also brought to light the problem of the weakening capacity of local governments to regulate and enforce social policies that protect both individuals and the environment from the power of global capital and institutions.

International financial and trade institutions were also targets of attention in Vancouver. Activists in Vancouver spoke of their concerns about the consequences of water privatization on local control of resources, in particular, on the ability for local governments to regulate water services in light of obligations under international trade agreements. The perceptions of risk expressed by activists in Vancouver centered on the financial and regulatory clauses contained in international trade agreements, including NAFTA (the North American Free Trade Agreement), that are used by foreign corporations to guarantee financial security from national governments.[9]

Both the BCPSEU and CAL used trade-risk themes to frame the problem of water privatization. Documents obtained from the two organizations, which were distributed at the public consultation meetings, consistently reference the potential risks to local democratic processes from NAFTA and other international financial regulatory bodies, including the WTO. One of the fact sheets on trade and water, created and distributed by CAL, provides examples of other cases in which foreign-owned corporations have demanded compensation for lost profit from the Canadian federal government to illustrate why water privatization was a financial and accountability risk for local governments. The document claims that "the privatization of water means future generations will have to pay billions in compensation just to govern our own resources."[10]

In public consultation meetings in the Vancouver region during the spring of 2001, activists from labor, social justice, and environmental organizations frequently used messaging about trade agreements and the threat to local sovereignty posed by outsourcing municipal services to foreign corporations. Karen Evans, an independent researcher and local organic farmer who has authored several publications on water and trade, attended all of the public consultation sessions on water privatization so that she could speak about the risks to local control under NAFTA. During my interview with her, she described herself as "one of the first Canadians to raise the issue of water seclusion in trade agreements" and expressed her deep concern about losing public control of water resources. Although she explained that in the past the issue of water and trade has generally not been raised in connection with privatization, she was amazed how many people "made that link" at the public meetings she attended. She described the sense of fear that permeated public discourse at the consultation sessions:

So many people raised this issue because I believe the Canadian public was scared. I think they have a very nebulous idea that there is some problem with trade agreements, and they fear walking down that road and then all of a sudden being sandbagged at the end. And of course it comes to sovereignty. I mean if there is any area where we need sovereignty, food and water are it. So I think the public was deeply concerned about NAFTA, which is so complicated and hard to understand, and the fact that local politicians simply did not get the risks associated with it.

Like others in the Vancouver movement, Karen Evans linked the fear of water privatization to national-level issues around concerns over trade agreements and the erosion of Canadian sovereignty.[11]

Beyond financial concerns, many activists in Vancouver made reference to the threats to ecological systems imposed by investor provisions under NAFTA and other international trade agreements. Their responses reflect an understanding of the altered relationship between capital, the state, and the environment in the world risk society, where global flows of capital result in unprecedented levels of resource depletion, waste production, and the corresponding social, economic, and environmental risks.[12] They described broader concerns about climate change and resource depletion. Many worried that the privatization of local water services would represent the slippery slope, leading to complete loss of water sovereignty in Canada. Peter Clark, a political activist who "come[s] very much from an environmental perspective," expressed his concern about the impact of trade agreements on the integrity of water systems: "Once we open up water to trade, Canada will just become the faucet to the north, and we will see a huge flow of water, which will be very detrimental to our environment here, especially with the impact of climate change." Beyond his fears about the loss of local control, Clark is concerned about the ecological repercussions of being beholden to trade agreements. "We know under NAFTA that once you trade a commodity you have to continue to supply that portion in perpetuity. And the real trouble with that is we don't know what climate change is going to bring in terms of shifting weather patterns that will affect the flow of water. Losing control of our water systems will be devastating to the ecology of place," he explained.

The use of legal expertise and channels is another strategy employed by movements to challenge the power of the state. Social movement organizations are increasingly using legal tactics to mobilize individuals, provide strategic resources, and frame arguments.[13] Vancouver's anti–water privatization movement was aware of this strategy and made use of it, albeit without resorting to litigation. In addition to the documents and fact sheets distributed to the public, the BCPSEU commissioned a legal document from an expert in international trade law, which was presented to elected officials and bureaucrats. The opinion was written by Clark Green, an expert in international trade law and partner in a prestigious law firm located in Toronto, Canada's financial center. He explained the dangers of Chapter 11, the investment clause under NAFTA, which, in his opinion, could be used in the case of water privatization to weaken the power of local governments to regulate water services. He said that the significance of trade and investment treaties in this case was that "ultimately Chapter 11 could seriously impair the ability of governments to make decisions around resources." He explained:

Well, Canada is a party to a number of these treaties and for the purposes of the Seymour plant, the most significant was NAFTA, Chapter 11, which protects investment. And dispute procedures are set out in the investment chapter, and they entitle foreign investors to file a claim for compensation for damages before international investment tribunals—in a broad, but ill-defined variety of circumstances in which governments do things that somehow interfere with the investment or the profits that the private investor was hoping to make from the investment.

Green made clear that the municipal and federal governments could have been financially liable for lost corporate profits in the case of a change in contract and drew attention to the fact that decisions about liability would not be decided in a Canadian court of law, but rather by a closed-door NAFTA tribunal, with no public consultation.

Clark Green explained that few people in Canada—including municipal government officials—have an expert understanding of how trade and investment treaties function and their ramifications for local control of public services and resources. The Vancouver campaign strategically used legal opinion to draw attention to the potential loss of local control and accountability by local governments over decisions about water resources. Highlighting the legal risks to the regional government under NAFTA added expert credibility to the movement's claims and worked to create opportunities favorable to its goals. "The NAFTA issue was the straw that broke the camel's back," Fiona Rogers told me, explaining that "[Metro Vancouver] basically ended up backtracking because they realized that once they went that route, they were setting a real precedent for the future control of water." The legal opinion, as opposed to the confrontational legal challenge utilized in Stockton, was valuable in opening up political opportunities for movement claims to be legitimized.

Using global risk frames to highlight the potential threats to local control under international trade structures was an effective mobilizing tool by the Vancouver movement, because it linked nebulous and complex international regulatory institutions to the decisions by local politicians who live and work in the same communities as their activist counterparts, creating a clear and accessible target. As Peter Clark explained, "The message 'focus on the municipal' is really powerful because [elected officials] know that you know where they live." Clark described that at the first public consultation meeting he "just walked to the front of the room and took the microphone from the chair of the water committee and said, 'This is our meeting and we are going to tell you what we want.'" He went on:

Then I said, "Is there anyone here," and there were well over five hundred people in the room; I said, "If there is anyone here who is in favor of privatizing our water, please stand." And nobody stood. And then I said, "Who here is opposed to it?" And everyone stood. And I said to the chair, "There is your consultation. You wanted consultation, you've got it. We don't want it privatized and if you privatize it, we are going to go after every single politician who is involved in this decision."

While activists in Vancouver drew attention to the global nature of the problem of water privatization, they also recognized the importance of local political process, including how decisions by municipal politicians can either exacerbate or limit the risks from transnational institutions.

By focusing on the loss of municipal control over resources and the fate of local elected officials in future elections, the movement was able to create openings for their claims to be considered. Local politicians and elite decision makers were concerned about the legal implications of international trade agreements—that are negotiated and signed without the input from municipalities—for communities that contract out services to the private sector. Many of the political decision makers in Vancouver discussed their concerns about the unknown implications of trade and investment treaties for local communities. Richard Martin, a senior Metro Vancouver bureaucrat, cited uncertainty about the impact of international trade treaties, including NAFTA, as critical to the decision by the water board to reverse plans to privatize water services in the region:

[W]e were legally vulnerable under NAFTA, and . . . we really hadn't looked into that enough. So we came back after two meetings, and I sat down with the chair of the water committee, who is a very big proponent of [privatization] and felt this was a really good idea, and I said to him, "It looks like we are going to be into a protracted political battle and a protracted legal battle. . . . This is clearly not just a local issue. This is . . . an international issue to do with international trade treaties. My advice, no matter what you think of [privatization], is that there is not enough potential benefit in this to keep on going in the face of this kind of legal and political opposition." And to my surprise he readily agreed. He, too, could see that this was just going to be the dog's breakfast in terms of the political process.

The legal argument by Vancouver activists reframed the issue in the minds of local politicians, shifting the focus from water privatization as an innovative idea, to the need for local governments to protect themselves in the face of powerful trade and investment treaties.

Despite initially feeling unreceptive to the anti–water privatization activists and their concerns, Richard Martin said that the movement's claims about the vulnerability of local governments in light of international trade treaties made him "take a close second look" at the decision to privatize. "Municipalities across Canada need to be in a better position to make decisions about resources, whether it is water or something else. They've got to know if they are safe or whether they should be hiring a battery of lawyers to protect themselves," he explained. Martin felt that by focusing on the risks from neoliberal corporate ideology and international trade agreements, the movement was conveying the message that they trusted public officials to oversee and regulate water, creating a sense of solidarity between local elites and activists.

Global frames—and the global connectors through which meanings were crystallized—were not only critical for movement mobilization, but also shaped movement outcomes in Vancouver. Global frames were successful in creating favorable conditions for the movement because they facilitated the coming together of elected officials and municipal bureaucrats with social movement actors, rather than pitting one against the other in a movement-countermovement interplay, as is often the case.[14]

Responses from activists in Vancouver demonstrate how global issues and international movement linkages create leaders who facilitate the flow of information and transform the nature of collective action at the local level. Transnational issues disrupt local-level politics and shape responses to political opportunities, revealing the interdependence of local-level social and political structures with global networks and flows of information. In Vancouver, global frames provided the inspiration and motivation to mobilize a wide range of constituents by connecting anti-globalization issues—including multinational corporate policies and the entrenchment of neoliberal policies—to a concrete local struggle and developing a sense of transnational unity with other communities through the use of global narratives. Global frames also acted as leverage to reframe the issue from a domestic political confrontation between local authorities and activists to an issue of global power and risk, creating a local solidarity that opened up favorable opportunities for a positive movement outcome.

Conclusion

The social and environmental consequences of neoliberal globalization are the subject of intense debate at the local level, reflected in the discursive struggles between those who ideologically support the expansion of the global economy and the opposition movements that oppose the increased intrusion of global capital into local social, environmental, and cultural domains.[15] The outcomes of these battles over frames are significant for whether or not neoliberal globalization will continue to encroach on local communities.

Although respondents in Stockton and Vancouver expressed their understandings of the importance of water as a public resource with equal conviction, the commons discourse was articulated into divergent arguments in each case—despite facing a similar threat of water system privatization. Comprehending the nature of the anti–water privatization movements in Stockton and Vancouver was linked to preexisting cultural, organizational, and political contexts. In each place, activists drew upon preexisting cultural codes, values, and norms, which shaped how the problem was understood and presented to political elites. The emphasis on democracy and voter rights in Stockton was tied to the litigious movement history of California—characterized by the use of voter-driven initiatives—and the political ideology of key movement leaders and organizations that emphasized the traditional left-right political divide. Most of the coalition leaders interviewed in Stockton were connected to organizations that were ideologically opposed to the local municipal government, and had very few ties to external organizations and individuals, including groups engaged in global water issues. By focusing on political accountability rather than the negative consequences of outsourcing water services to a multinational corporation, the movement emphasized the division between activists and political authorities, and constrained their ability to counter the pro-privatization frames of the local government.

On the other hand, in Vancouver, the broader cultural and organizational context converged around issues of social justice and globalization. Key movement leaders and organizations (that I call *global connectors*) were linked structurally—through network ties and previous involvement in the anti-globalization movement—and cognitively—through preexisting understandings of the consequences of globalization—to transnational water activists and institutions that shaped the construction of global risk frames. The movement's focus on the power of capital and trade institutions to alter and threaten local decision-making capacity resonated with political elites, drawing attention to their legal vulnerability under NAFTA and demonstrating the common fate shared by all local citizens.

Frame processes are fraught with negotiation and conflict as movements attempt to synthesize and represent competing visions of reality. Movements that mediate differences and construct arguments that are concordant with political elites are more successful in presenting well-articulated, unified arguments that resonate widely and create favorable opportunities.[16] In Stockton, the movement failed to negotiate and present clearly defined and well-articulated frames, instead focusing on complex and divisive issues of political corruption, democracy, and voter rights. The organizational culture and historic movement focus on local political disputes shifted attention away from the global problem of water privatization, emphasizing the schism between movement actors and political authorities, and thus failing to neutralize pro-privatization frames. As a result, the legitimacy of the movement's claims was rejected by decision makers, constraining their ability to influence the city council's decision on privatization.

In Vancouver, on the other hand, organizational ties and previous experiences with anti-globalization issues linked activists at the local level to a network of global activists and to broader frames of transnational environmental and social justice. The strategic use of anti-globalization discourse created a sense of solidarity in the face of global foes, and moved the struggle away from the activist/elite dichotomy reflective of many social movements. By constructing frames around global issues, the movement in Vancouver was able to present a clearly defined problem— the international risk to local control of water—which, in turn, led to a solution-oriented frame—keep water in public hands—that resonated with political elites and neutralized pro-privatization arguments.

Differences in the integration of transnational concerns into claims-making processes across the two movements highlight the importance of *global frames* and *global connectors* to local movements implicated in transnational flows of capital, institutions, and trade regulations. The use of frames that take into account the changing nature of social and environmental risk and institutional power—from local or nation-specific contexts to the transnational realm—are more likely to be favorable in opening up political opportunities than those that focus uniquely on local issues and grievances. Because anti–water privatization movements are connected to wider global processes, including multinational corporate policies and transnational financial institutions, local activists who incorporate global frames into their repertoires of contention can potentially mobilize a wider pool of supporters and present arguments that resonate with political authorities. By targeting global economic structures and institutions, activists create opportunities for shaping policies on local resources and services.

In Stockton, drawing attention to domestic political grievances exposed the schism between activists and authorities and worked to alienate decision makers. In Vancouver, the focus on global problems drew blame away from the local politicians and provided them with an "escape route" to reverse their decision without reflecting negatively on themselves. Global frames are important for local movements implicated in global processes, but local context also matters for the strength and durability of these movements because individuals are motivated by issues that have direct resonance in their daily lives. The resonance of global frames for local movements is contingent upon the presence of locally embedded activists—similar to Tarrow's "rooted cosmopolitans"—whose networks, collective identities, and shared understandings are connected to wider global struggles, and who are able to bring global issues down to the local level.[17]

Linking the problem of water privatization to global risk frames is strategically important for local anti–water privatization movements because they move the struggle beyond the conflict between local activists and political elites and into the international arena, revealing the serious threats from institutions beyond the control of locally elected officials. Drawing attention to global risks creates local leverage to legitimize movement claims and bring together social movement actors with decision makers. Frames that connect the local to the global are more readily received by domestic political power structures because they illuminate the threat to local control and accountability as well as create the sense of solidarity between local citizens and their governments in the face of global risks and power structures. By introducing and emphasizing a new angle—*global risks* and *transnational power structures*—local social movement actors can create new opportunities at the institutional level and increase their ability to shape policies relating to resources and services.

5

The Political Process: Seizing Local and Global Opportunities

Why did they fail? That's a good question. I know that specifically there was a big controversy whether or not to hire petition gatherers. Some people felt that they needed to do everything grassroots, with volunteers. They thought it would make them seem better than the mayor, who was really underhanded in his tricks. But in the end they didn't get enough signatures for the referendum . . . they fell short because they decided not to hire petition gatherers. I think it might have made a difference. But, you know, in short, the crude answer is they didn't build enough power. They simply didn't demonstrate enough opposition to the contract.

—Craig Butler, community organizer, Stockton

There are so many reasons why water issues need to be part of a global movement and a local movement. As water becomes more scarce, it's going to become a massive issue on a global level. I think we actually should be proactive about setting in place policies that support the notion that water is ours to protect, and keep under democratic control. We definitely pointed to that global connection and the fact that as demand increases, there will be more and more pressure for someone to make a profit from selling water. People were really fired up about that, and they knew how to disrupt those meetings and really make it an unfriendly environment for the Metro Vancouver crew.

—Fiona Rogers, community activist, Vancouver

In Stockton, the closed nature of local political structures constrained the movement's ability to access elites and influence policies around water services. The movement activists also lacked experience at targeting local political structures and downplayed (intentionally or not) international threats. The result was a failure to shape political decision making. Yet, ultimately, through legal challenges, the movement in Stockton overturned the private contract and returned control of water services to the public sector. This reversal represents an important and significant victory and should send a signal to social movements about the role of legal opportunities for shaping outcomes.

In Vancouver, greater institutional access, enabled in part by the de-centralized decision-making structure of the Metro Vancouver region, provided openings for movement actors to contest the proposal for priva-tization and create cleavages between political elites. Preexisting move-ments, particularly those focused on watershed issues, opened up political spaces for public deliberation and fostered early political alliances. The intertwining of global and local opportunities also created new political openings that allowed anti–water privatization activists in Vancouver to influence policy.

Movement building is dependent on the conditions of the broader political context. Beyond cognitive strategies such as frames, political opportunity structures shaped the diverse trajectories of the anti-water privatization movements in Stockton and Vancouver. Despite facing similar overall circumstances—the proposed privatization of local water treatment and delivery systems by multinational water firms and an over-arching national context of neoliberal policy reform—the movements in both places emerged and responded differently. The entrenchment of economic globalization has, to some extent, diminished the authoritative power of states to regulate and control resources within their boundaries and make decisions on behalf of their citizens; at the same time, it has constrained the ability of social movements to shape domestic policies.[1] Yet, the evidence from the Stockton and Vancouver cases suggests that local political structures continue to matter for the trajectories and out-comes of mobilization.

Anti–water privatization movements are influenced by processes oc-curring at multiple spatial levels, including the local, national, and in-ternational political economy. While water is a geographically bounded resource that is regulated by local and regional governments, the expan-sion of water privatization globally means local water systems and ser-vices are increasingly influenced by the actions of global institutions, including multinational water firms that bypass the nation-state to directly target municipal governments. What are the implications of the state's di-minished capacity and legitimacy in regulating environmental protection in the face of increased economic deregulation? How are movements re-sponding to this shift in power from the state to non-state institutions, such as corporations and international financial bodies? How do local movements, including anti–water privatization movements, disrupt flows of neoliberal corporate power by seizing international opportunities and reconstituting them in ways that resonate with domestic political oppor-tunity structures?

This chapter examines political conditions in Stockton and Vancouver and reveals differences across three levels: the degree of institutional openness, the nature of preexisting movements, and the way in which opportunities were seized by activists, including those at the local and global level. Beyond the structural conditions of the polity, the actions of social movement actors matter to movement mobilization and outcomes. Movements that seize both local and international opportunities and target their grievances accordingly are more likely to be successful in their outcomes. By drawing attention to international constraints on the decision-making and regulatory capacity of municipal governments, anti–water privatization activists open up opportunities for grievances to be considered.

The Changing Role of the State: Local Governments and the Urban Water Crisis

While government institutions, political culture, and the predominant economic sectors in Stockton are significantly different from those in Vancouver[2], both communities have been shaped by similarities in the broader national and international context, characterized by the downward shift—from national to local—in regulation and service delivery. The federal disinvestment in infrastructure and services that has occurred since the 1980s in both the United States and Canada coupled with increased competition for water resources—from expanding urban populations and demands from the industrial and agriculture sectors—has resulted in an urban water crisis. With the shift in responsibility for water governance from federal to municipal governments in a neoliberal era, local governments are struggling to meet the challenge of providing clean, high quality drinking water, managing resources efficiently, and implementing conservation measures.[3]

Public water systems in North America and internationally require enormous investments in new technologies in order to address crumbling infrastructure, deal with scarcity, and ensure a clean and adequate supply of drinking water for current and future generations.[4] How are local utilities adapting to new demands for increased investment in water services and infrastructure upgrades that will enhance water quality and guarantee stewardship? In an era of fiscal austerity, many governments have turned to the private sector to finance new water utility projects, increase efficiency in management and delivery, and implement conservation measures. Since the 1990s, multinational water companies have been aggressively lobbying local governments to privatize their water services, arguing that the private sector can manage and deliver water services more efficiently.[5]

Most local governments are outmatched by large water conglomerates, which have access to the legal, financial, and marketing resources needed to convince decision makers to outsource water services. As Jeffrey Roth-feder explains in his book *Every Drop for Sale*:

Water privateers use an encircling technique. They prepare detailed reports and proposals that are steeped heavily with statistics and conclusions supporting the premise that municipal control has been a failure and proposing to supply water for the local government at a much cheaper price. Then they curry favor with the local politicians through contributions, dinners, and tickets to events like the Super Bowl or the World Cup. In European countries, where bribing local officials is more common than in the United States, money changes hands as well. Then they deliver the presentation.[6]

This is the new reality for local governments around the world, and not just those in poor or developing countries. It is easy to see why it is difficult for governments to say no.

Facing the need to invest in local water utilities to meet new environmental regulations and increased competition for water resources, and lacking the financial capital to meet these demands, the municipal governments in both Stockton and Vancouver felt they had no choice but to outsource water services to the private sector, setting the stage for a battle with their citizens.

Stockton: Closed Structures and Conservative Tactics

In Stockton, anti–water privatization activists were confronted by political institutional closure that constrained their ability to have their claims considered or form alliances with political elites. In 2002, a broader national- and state-level conservative agenda of small government, deregulation, and privatization, coupled with federal cutbacks to investment in infrastructure and services, as well as shrinking city budgets, shaped the Stockton City Council's decision to privatize municipal water services.[7] Federal cutbacks have had an enormous impact at the local level where municipalities are forced to grapple with fewer resources to pay for critical infrastructure upgrades, including water and sewage works. Fiscal restraint combined with neoconservative politics created the conditions for municipal governments to embrace market reforms at the local level.[8] The entrenchment of neoliberal policies in the United States that began in the 1970s and expanded rapidly during 1980s and 1990s, under both Republican and Democratic administrations, also provided openings for the involvement of the private sector in the delivery of public services, including the expansion of public- private partnerships at the municipal level.[9] Multinational water companies have capitalized on these market reforms by lobbying local governments for private infrastructure contracts for drinking water and sewage treatment services. This broader national and local conservative agenda set the stage for the proposal to outsource water treatment in Stockton.

At the time of the proposal to privatize water services, the mayor of Stockton had recently returned from the U.S. Conference of Mayors, where he met with representatives from private water firms vying for control of Stockton water services. When the Stockton City Council was required to upgrade the municipal water treatment plant, the mayor turned to the private sector to finance these upgrades, and oversee the delivery of municipality water services. As a prominent and successful business leader in the community, the mayor believed in running the city on a business model, and was convinced that the private sector could deliver services more efficiently and at less cost to the taxpayers.[10]

Barbara Smith, a sixty-eight-year-old grant writer and former teacher, and a member of the anti–water privatization coalition's steering committee, linked the push for water privatization to the ideology of the mayor at the time. "The city council had a conservative platform, driven by the mayor, who was very popular at that particular point in time. And people really wanted to listen to him," she said. Smith went on to explain that people admired the mayor because of his business acumen and felt that by "bringing a business model to the city," he would save the city money. "Because heaven knows government is wasteful!" she exclaimed, laughing. She pointed to the conservative national climate at the time as driving the push for water privatization in Stockton, noting: "Some of the actions by Bush and Congress favored privatizing. And they had all the usual rationale. He [the mayor] had gone to the mayor's conference where they promote privatization for infrastructure upgrades and came away feeling that this was a way of the future." The right-wing ideological stance of the mayor and the majority of the city council, strengthened by a national Republican agenda that demonized government and favored the private sector, was described by many of the activists interviewed in Stockton as the driving force toward water privatization.

Beyond political ideology, activists in Stockton located the push for privatization in the agenda and influence of multinational corporations that target municipal politicians to secure lucrative infrastructure contracts. Graham Davis, an elected representative in San Joaquin County, explained that what happened in Stockton was part of a larger trend across the United States: "You had the push from these big international water companies going through Minneapolis-St. Paul and . . . Atlanta and many other cities in the United States. And it became pretty much a good Republican agenda throughout the country. So essentially with all those things . . . it just got itself alive here in Stockton." Davis told me that the mayor pushed the idea that the private sector is more efficient and effective, in part because he had been influenced by executives from major multinational water corporations at the national mayor's conference he attended prior to the privatization proposal: "These companies told [the mayor] that it's always cheaper to privatize, and private industry always operates higher quality. So he came back and said that that's just the way it is, and he just kept pushing this thing and pushing this thing." Having attended similar conferences in the past, Davis described being "wined and dined" by private sector firms looking to bid on municipal infrastructure projects. Eventually, the mayor became "very forceful in his position in trying to privatize this issue and through whatever means he was able to generate enough votes to move it forward." A belief in the superiority of the private sector to maximize efficiency and lower costs as well as the lobbying influence of multinational corporations undergirded the pro-privatization position of Stockton's mayor and city council.

Many activists in Stockton described the conservative stance of the council majority as preventing the movement from accessing the necessary political channels for presenting their arguments against privatization. Some respondents even argued that the closure of political structures went beyond ideology, and reflected the general atmosphere of secrecy that characterized the city council at the time. Kelly Jones, a forty-five-year-old environmental manager for the city of Stockton, described the lack of transparency of the municipal government as shaping its refusal to consider the claims of the anti–water privatization movement. "Well, the mayor and council certainly had their agenda. . . . They decided that this is what was going to happen, and they weren't going to hear anything else. Part of the whole problem with the way local politics have played in this town for decades and decades is that decisions are made behind closed doors . . . with no public transparency. And so when that whole decision came down it was just more of the same," she said. Despite attempts by the movement to have their concerns about privatization considered by the city council, including attending council meetings and making presentations to council, a closed-door policy on the part of the local government prevented the movement from having their claims heard.

Even city officials opposed to privatization were met with hostile responses from the mayor and a majority of the city council when they attempted to present arguments in favor of maintaining a publicly run water system. Paul Anderson, a former employee of the city of Stockton, who worked in the city manager's office at the time of the fight over privatization, was opposed to outsourcing the city's water treatment and delivery services on the grounds that it would be less cost-effective than keeping the services in-house. When he heard about the proposal to privatize the water treatment plant, he quickly mobilized employees in his division to complete a comprehensive cost-benefit analysis of the water treatment plant, to demonstrate the advantages of maintaining a publicly run system. Despite the evidence showing that keeping water services in the public hands, albeit with considerable restructuring, would be less costly to taxpayers than privatization, Anderson was met with skepticism by the mayor. He described the reaction of the mayor to the business plan created by the city manager's office:

Even with the plan and the cost-benefit analysis, essentially the mayor's attitude was, "I really don't care about any of that. We just need to privatize because they can do it cheaper." It was a little bit like George Bush and Iraq. Every time you cornered him, he changed his reason for why to privatize. In the end, he got the majority of the council to agree to go out and bid on a private contract. The mayor had effective psychological control over a majority of the council, regardless of the issue, and he simply drew that line. He knew that he wasn't going to convince the minority, so he just moved with his majority.

Despite the fact that the city staff was opposed to privatization and presented well-researched and documented evidence about the higher costs associated with privatization, the mayor remained entrenched in his pro-privatization position, and used his persuasive ability to convince other elected officials to ignore the counterevidence presented by bureaucratic insiders and movement activists.

The relatively closed nature of the local government structures caused activists to turn their attention away from alliance building with political elites and downplay anti-corporate messages so that they could concentrate on local democratic processes, including tactics aimed at usurping the power of the council through political and legal challenges that would block or delay privatization. Although experts on water privatization and activists from outside of Stockton emphasized the importance of global issues, including the implications and risks of privatization under NAFTA and the negative track record of multinational water companies in other communities, the movement in Stockton chose to focus on legislative challenges.

When movement leaders realized they were being stonewalled by the city council, they decided to push for a ballot initiative requiring voter approval for any private contracts for municipal services over $5 million, including water and wastewater treatment. "And we won that initiative by a 60 percent margin. But, as you know, the city council went ahead and signed the contract before the vote came in, and it is not retroactive," lamented Barbara Smith. She went on to explain that the movement's subsequent attempt to organize a referendum to overturn the council's vote was also unsuccessful because they were unable to gather the required number of signatures. Their efforts were further hampered when the mayor became personally involved in trying to prevent the movement from securing the required number of signatures for a referendum vote. "He started calling the citizens of Stockton asking them not to sign the referendum petition," said Smith, adding that a private company was circulating a counter-petition and the mayor tried to "convince people to sign the counter-initiative which would effectively remove their name from our list if they had already signed." She described how the mayor "basically told them that we were lying and misrepresenting the city council."

The counterattack by the mayor and council was in part due to the movement's own negative attacks on locally elected representatives and their decision-making procedures. Some of the people interviewed in Stockton felt that the constant vilification of the mayor was harmful to the movement's chances of preventing privatization because it made authorities unreceptive to their arguments. Tom Bailey, a local business leader and proponent of privatization, explained that the antigovernment stance by anti–water privatization activists caused the mayor and some of the members of council to entrench themselves further in their pro-privatization position. "They really felt that they had been elected to lead and make decisions on the behalf of the citizens of Stockton who had voted them into office. And the mayor in particular felt that his integrity had been attacked," Bailey explained, adding, "you know, why should he have to have a public referendum on every decision being made when he was elected to make those decisions on behalf of the public?" According to some activists in Stockton, by maligning the decisions of the mayor and city council, the anti–water privatization movement reduced its chances of success by strengthening the solidarity among political elites.

Others felt that it would have been strategic to target local elected representatives and create openings for movement claims to be considered by emphasizing the threats from international institutions and multinational corporations. They felt that they might have captured the attention of local officials by illuminating the threat to local control from global institutions. Some of the anti–water privatization activists in Stockton emphasized the importance of global processes for shaping decisions about the delivery of local water resources, including the implications for local regulatory control under NAFTA and the threat from multinational corporate bottom-line policies. Yet the focus on international threats and opportunities never became central to the campaign. Barbara Smith thought that the movement should have concentrated more on global concerns to raise awareness of these issues with the Stockton City Council: "Some people from the university and also people from [the Citizens Environmental Alliance] came and made several presentations on NAFTA and the potential consequences if things got sticky down the road. They also talked about how risky it was to trust these contractors and their spokespeople because their interests are only financial—they are looking to make a profit and not to strengthen the community." She thought that making connections to global issues and highlighting concerns about NAFTA would have helped mobilize the broader community and "would be useful to bring to the council," but that "the movement didn't end up using those arguments. . . . I am not sure that was such a good idea in retrospect."

The inability of the movement to create institutional openings and build alliances with elites was exacerbated by the relative inexperience of anti–water privatization activists at targeting local government. Paul Anderson described the lack of local political organizing experience on the part of anti–water privatization activists:

Many of the people had experience with advocacy and organizing, but the focus was always on state or federal policies or statutes. There weren't many people involved who had strong connections or experience with how local government works. The coalition came together for the first time around water privatization. And despite their best efforts they never really built the kind of grassroots movement needed to stop [the mayor] from pushing through [privatization].

This meant that few of the people involved in the movement in Stockton had ties to municipal authorities.

Some of the respondents in Stockton pointed to the inexperience of activists in local politics as leading to errors in tactical decisions that resulted in the movement falling short of its goals. Bernie Jacobs, a retired professional and member of the anti–water privatization coalition steering committee, explained that the decision to use volunteers rather than professional signature gatherers in a campaign to force a referendum was one of these instances: "I don't know who or what or why but somebody decided and somebody agreed with [the decision] that we should not use the professional signature collectors." He explained that to get a referendum to overturn the city council vote, they needed to gather the required number of signatures in thirty days. "I thought we are going to be able to do this since we had already won the [initiative] vote by 60 percent. And then I learned all we are going to do is a weekend of door-to-door signature collection. I couldn't believe it. I mean how we were going to do all of that in four or five weekends when it took us three months of professional signature collection daily to collect enough signatures for the initiative? It was absurd!" he exclaimed. Jacobs argued that this demonstrated a "lack of understanding about community organizing" by movement leaders in Stockton.

Bernie Jacobs blamed the inexperience of the steering committee members for strategic missteps that allowed the council to proceed with water privatization. He described how he felt when the movement failed to gather enough signatures for the referendum: "Oh, it was awful, awful! This was the big disappointment. You know it is so upsetting thinking about it, because had we [used professional signature gatherers], then privatization would have been stopped in its tracks. It wouldn't even have begun."

Other members of the movement in Stockton shared Bernie Jacob's view. Charles Barlow, a fifty-two-year-old lawyer and environmental activist, explained how he felt:

The steering committee was politically naive. They had never done anything like this before and they felt that it could be done only on weekends. We were unsuccessful because there was a strategic blunder, which was not realizing that using volunteers for collecting signatures only goes so far and when you have to collect 15,000 signatures over four weekends! You can't do that without putting out the bucks and bringing in the professionals. And we would have won and stopped privatization if we had been more strategic. . . . And we would not have had to go through that whole surge of litigation and all the costs associated with it.

Beyond inexperience and missteps, the choice of targets and the tactical repertoire adopted by the movement in Stockton were also critical for influencing movement outcomes. Although some activists in Stockton attributed the referendum failure to a general lack of experience on the part of many of movement activists, others said that the decision not to use professional signature gatherers was a strategic choice on the part of the movements to counter what they considered to be underhanded behavior of the mayor and city council. Joan Davidson, a former elected representative and member of the anti–water privatization coalition explained that part of the reason for not using professional signature gatherers was the belief on the part of some coalition members that they were taking the political "high road" by using grassroots volunteers and the feeling that "because they felt they were right, that somehow things would work out."

By taking the "high road," activists in Stockton also deliberately chose more conservative tactics in presenting their case against privatization. Many respondents said that the movement wanted to avoid appearing emotional and angry lest they be accused of lacking the knowledge and expertise to counter pro-privatization arguments. Despite a push by some of the anti–water privatization coalition members, including youth and union representatives, to engage in more disruptive protest tactics, the coalition steering committee ultimately decided to take a more moderate tactical approach by focusing on evidence-based arguments.

Bruce Owen, a retired professional and member of the anti–water privatization steering committee, was opposed to the use of disruptive protest tactics by the movement. "The radical elements of certain movements, like some of the environmental movement, give the rest of us a bad name," and diminishes the movement's effectiveness, he argued. He has held leadership positions in several prominent environmental organizations and has been a member of the environmental movement for over twenty years, describing himself as a lifelong conservative who supported Republican candidates for most of his life. He believes in private enterprise and that water privatization is the right solution for many communities that lack the infrastructure needed for high quality water services, even if it in not the right solution for Stockton.

"Candlelight vigils" were not the way to fight privatization, but rather, "research and analysis and informed processes," according to Owen, who explained:

The opposition, the city council, and their attorneys tried to paint us as emotional. So we felt we had to fight being perceived as emotional. In the initial stages, we had some wonderful people, but they basically wanted to hold a candlelight vigil and protest on the steps of city hall. Well, they would love us to hold a thousand candlelight vigils and protests. While we are spending time doing that, they are off and they are doing their thing. When you are focused on those kinds of things the other guys are moving the process along at a rapid enough pace that you can't catch up. So you need to play the facts-based game.

In an attempt to appear professional and unemotional, the steering committee focused their attention on less disruptive tactics.

Other activists involved in the anti–water privatization movement felt that the moderate tactics failed to mobilize a broad-based opposition movement. Scott Winslow, a union organizer, joined the anti–water privatization movement in Stockton because he felt that a bad policy decision was being railroaded through the community by the mayor and his supporters on the city council. He disagreed with the decision by the steering committee to use less disruptive tactics, and felt that a more aggressive stance would have mobilized the broader community, and made the council pay attention to the movement's concerns: "We should have done something more than just make nice presentations to the council. We should have said, 'Hey boys, you work for us and your instructions are this.' And we should have locked that goddamn place down and—I get a bit warmed up—we should have said, 'You work for us and you need to know it.' And we should have made goddamn sure at every chance we had to know it!" Many respondents echoed similar concerns about the tactical choices of the coalition steering committee, and thought that a more aggressive opposition movement would have contributed to a successful outcome.

Paul Anderson said he felt that the decision to play a "facts-based game," rather than mobilizing broad community support, resulted in the failure to block water privatization. "They were doomed to lose with their facts-based approach. Had they doubled up and ran a 'You dirty rats,' a really strong 'You dirty rats, you are violating democratic principles here' campaign and presented a more angry face for the movement, things might have been different," he said. Anderson felt that the movement should have "focused more attention on grassroots organizing" in order to "build a powerful enough opposition." He explained that there were many youth activists who "wanted to organize demonstrations and generally be loud and use in-your-face kind of tactics," and told me that the movement leaders dismissed the use of protest tactics because they felt they had to take the moral high road in response to the actions of the mayor and council. Anderson felt that grassroots organizing and boisterous protest tactics were their only chance of securing a win: "If they had been able to show how widespread the opposition was . . . then the issue would have died. . . . That was their only chance to win. Otherwise they were doomed to lose." When asked why he thought the movement failed to pursue a more aggressive organizing model, he said: "Because [the steering committee members] are facts-based people and I don't think they really understood the whole [idea of] what it takes to win. You know, if Saul Alinsky [the legendary community organizer] had been in town it would have been a whole different game. They might have stood a chance at winning."

Overturning Privatization: The Role of Legal Opportunity Structures
While the anti-privatization movement in Stockton was unable to prevent the privatization of the water treatment plant, they were ultimately successful in overturning the private contract that was awarded to OMI Thames in 2003. Rather than simply give up in the face of defeat, activists

in Stockton mounted a legal challenge to the outsourcing of the water treatment plant, and argued that the city of Stockton violated the California Environmental Quality Act (CEQA) by not requiring a full environmental assessment of the proposed infrastructure upgrades to the municipal wastewater utility plant. With support from a team of lawyers with expertise in environmental law (from a prestigious law firm based in San Francisco), the anti–water privatization coalition challenged the city of Stockton and OMI Thames in court. After a drawn-out process, with wins and appeals on both sides, the coalition prevailed when city officials decided not to appeal the final court decision that ruled against the city of Stockton and OMI Thames. The private contract was rescinded in 2008, and the water treatment plant was turned back over to the public sector. In an ironic twist, the city of Stockton was unable, even without the legal appeal, to renegotiate a contract with OMI Thames that would meet the requirements of CEQA because of the successful ballot initiative (Measure F) brought about by the anti–water privatization movement, requiring voter approval on private contracts over $5 million. Any new contract would have to be approved by the citizens of Stockton, an impossible task for elected officials in Stockton under Measure F.

While many of the activists interviewed felt that the anti–water privatization movement did not fail in its efforts to block privatization because of the eventual legal outcome in favor of the coalition, others described the win as pyrrhic victory because of the devastating cost to the community from privatization. Dave Alexis, a fifty-six-year-old business owner and former elected representative, explained his feelings:

For me [the legal win] was bittersweet. I mean it was good that the right thing ended up happening, and the truth came out, and that it showed we can't circumvent process. The bitter part was that we impacted human lives, and that this experiment cost the city a lot of money, and it was a huge cost in terms of liability, too. And there is still one employee who has fallen through the cracks and is unemployed and has serious lifetime medical issues, whose life is now in limbo. I felt vindicated [the way you do] when you stand up to something but also very bad in terms of the costs for our citizens and these groups and the city, and the cost in terms of dollars—but more important, the human cost, which is almost irreparable.

The high financial costs, job losses, and human suffering that resulted from privatization cast a shadow on the subsequent legal victory for some of the activists in Stockton. Bernie Jacobs lamented the financial costs to the taxpayers of Stockton: "If we hadn't made those mistakes, [there wouldn't have been] the several million dollars in costs that privatization brought about plus the millions it cost to transition back to the public.

That's a huge waste of money based on mistakes. It is important to know it to avoid making those same mistakes again." Although they ultimately prevailed against the proponents of privatization, many people who were closely involved in the movement expressed mixed feelings about the legal win; while positive about the return of the water system to public control, they lamented the personal and financial costs associated with privatization and the costs incurred by the city as a result of the legal challenge that ultimately reversed privatization. They were also disappointed in their failure to stop privatization in the first place.

In Stockton, the closed political institutional structures, combined with the centralized power and ideological stance of the mayor and the majority of city council members, prevented the anti–water privatization movement from successfully blocking the privatization of the municipal water treatment plant. The lack of preexisting movements focused on local electoral politics, as well as the relative inexperience of the activists involved in the fight against privatization, prevented the movement from forming alliances with elites, and gaining access to internal political structures. As a result, they focused more attention on targeting the mayor and council, and downplayed the use of disruptive protest tactics and linking the issue to global concerns, thus failing to mobilize the kind of broad community activism and engagement needed for a successful outcome.

Despite initially failing to block privatization, the success of the legal challenge launched by Stockton activists ultimately overturned the privatization of the water treatment plant demonstrating that even when faced with closed institutional access, movements that draw on legal tactics can force openings at the political institutional level. This is especially true in the United States, which is highly receptive to legal challenges to policy decisions.[11] In California, as in many regions across the United States, citizens often use the state's initiative process—a form of direct democracy, which allows citizens to draft laws or amendments and place them on the ballot—to make policy decisions. In the United States, and in California in particular, citizens and organizations have been highly successful in using the initiative process to enact new laws.[12] At the local level, initiatives are most often used to implement new growth-management policies, including urban growth boundaries to curb sprawl.[13] City-level initiatives most often occur in large, expanding, economically diverse urban areas and tend to focus on land use and the environment, including development and transportation. Those that achieve approval by voters are most likely to concern water, public utilities, and taxes. Local initiatives are particularly robust in California and are more likely than those at the state level to be enacted into law.[14]

The eventual success of the anti–water privatization movement in Stockton reveals the importance of the role of extra-institutional legal opportunities—including the ballot initiative process—in shaping the outcomes of social movements and demonstrates the need to adopt a broader understanding of movement outcomes by looking beyond the success or failure of movements in achieving their immediate goals.[15] Examining the consequences of sustained mobilization—across different stages in time—allows for a more comprehensive understanding of the impact of social movements, how they shape broader policy changes, and how success is measured.[16] In the end, the Stockton activists successfully reversed privatization and created the conditions for protecting water as a public resource in perpetuity.

Vancouver: Institutional Openness and Targeted Structures

In contrast to Stockton's relatively closed and unreceptive political structures, Vancouver activists faced favorable political opportunity structures and institutional openings, despite an overarching context of fiscal conservatism and private sector expansion in the public arena, similar to Stockton. The decision to consider outsourcing water services was shaped by a broader national context of neoliberal policy reform. A shift toward neoliberalism in Canada began in the 1980s and rose to prominence in the 1990s, when the governing Liberal Party focused on reforming the traditional role of government as the provider of services by engaging in a program of deregulation, tax cuts, and disinvestment in public services.[17] These market-based reforms set the stage for increased private sector involvement in public service provision, including water treatment and delivery.[18] The internationalization of financial capital and the negotiation and implementation of NAFTA also created the conditions for the increased privatization of services by emphasizing economic deregulation and removing barriers to foreign investment in Canada.[19] With the growth of multinational water companies, cash-strapped municipalities in Canada have been increasingly targeted by global water firms seeking to invest in lucrative water infrastructure projects.[20] This wider national and international context not only transformed the political economy of the federal and provincial governments, but also led to new orderings at the municipal level, including the embrace of market reforms and the adoption of a business model by local governments.[21] In a broader national context, neoliberalism contributed to the initial decision by the Metro Vancouver Board to privatize the Seymour water filtration plant and contract out water service delivery. This context also provided the opportunity for a coalition of anti–water privatization activists, political leaders, and organizations to mobilize against water privatization.

At the time of proposed water privatization in the Vancouver region, the Metro Vancouver Board was dominated by politicians from right-of-center parties in the region who favored the privatization approach. Many of the activists and opponents to privatization described what they perceived to be the ideological beliefs of some Metro Vancouver representatives—including the notion that the private sector is more economical and efficient at delivering services than the government—as driving the proposal to privatize water services.

Mark Spencer, an elected representative in the region, has been involved in politics for over twenty years, and has twice served as a municipal representative on the Metro Vancouver Board. At the time of the board's proposal to privatize the filtration plant, he had recently been reappointed to the board, and was shocked to hear that discussions about outsourcing the region's water services had been ongoing since 1996. Although he was dismayed by the initial decision to consider outsourcing water treatment and delivery in the region, he was not surprised: "In essence, the right wing pretty much had a chokehold on the Metro Vancouver government, and it was clearly their agenda to include a P3 [public-private partnership] project. It is like a religion to them. I think they view this much like Catholics view edicts from the pope. It doesn't have to make sense. You have to believe." He continued: "So anybody who tells you that it was about pragmatics is misleading you on the issue because for the people on the right, it was purely a belief that the private sector does everything better. And their refusal to see otherwise made a lot of people angry."

Spencer described the unwillingness of some Metro Vancouver board members to examine the pros and cons of water privatization as a mobilizing force for anti–water privatization activists. He felt strongly that because privatization is an ideological issue for politicians representing both the right and left of the political spectrum, the decision to privatize water services in the region should not be made by "a few bureaucrats and politicians," but should include input from the public. He used his role on the Metro Vancouver Board to mobilize the opposition movement and counter the pro-privatization arguments.

Despite the pro-privatization position of many Metro Vancouver representatives, the movement in Vancouver was able to overcome these constraints and create openings for their grievances to be heard. One reason for this success stemmed from activists' efforts in forming alliances with elites. Anti–water privatization activists in Vancouver strategically targeted elected politicians in municipalities across the Metro Vancouver region, who they felt would support their claims. Jim Roberts, president of the BCPSEU, the union at the center of the anti–water privatization movement, and one of the key leaders of the coalition, described the strategy: "Well, one of the things we did was to target politicians at the Metro Vancouver level. We felt there were many progressive councilors who could speak out at that level and make a difference. And many of them supported us. . . . We had some councilors who were actually coming out publicly and saying that we need to rethink this. That worked hugely in our favor." His response reveals the importance of establishing ties with elites to create openings for movement claims to be considered.

As part of his union organizing work, Jim Roberts frequently seeks support from sympathetic elected representatives, and explained that, in his opinion, campaigns targeted at influencing public policy are rarely successful without this critical support from political authorities. The strategic alliances between anti–water privatization activists and elites in the Metro Vancouver region allowed the movement's claims to be represented at the decision-making level. These alliances also created cleavages between authorities, with some aligning themselves with the movement opposing water privatization, while others remained entrenched in a pro-privatization position. Preexisting network ties between social movement actors and alliances within decision-making circles facilitated the movement's ability to generate support from political leaders.

In Vancouver, many of those who became active in the movement were trained in advocacy work, and there were others, both activists and politicians, with previous ties to environmental, social justice, and labor organizations, who formed the anti–water privatization coalition. Sherry Carruthers, a youth activist, explained that many of the people involved in the movement to stop water privatization in Vancouver had been trained in how to target and build relationships with politicians. During the campaign to stop water privatization, she had been working with the Center for Global Justice, an organization that trains young people to become leaders in their communities and globally, on campaigns designed to mobilize youth and provide them with the organizing and communication skills necessary to lead successful campaigns. As a twenty-two-year-old, Carruthers said she felt nervous about confronting politicians, yet at the same time, she felt prepared for these meetings because of her previous training with the Center for Global Justice, where she "was really involved with advocacy work." She went on to explain:

Those of us involved weren't scared at all, even though we were a youth group, with making connections and schmoozing and charming politicians at any level. A lot of the people involved had been involved with other campaigns, against the MAI in 1998 and the protest in Seattle in 1999. And young people had been trained by the labor movement and the environmental movement about how to go about grassroots organizing. And so we were quite adept at lobbying, at finding people to connect with at provincial and municipal levels.

Formal training and previous experiences targeting politicians facilitated the connection and support with elites used by respondents in Vancouver to bolster their cause. Activists lobbied city councilors, attended meetings of public officials, and sent information packages to elected officials in the region.

Along with advocacy training, the political openings and preexisting ties with authorities created by previous movements also facilitated public input on water privatization in Vancouver. In particular, prior movements focused on watershed protection had become institutionalized after decades of protest and advocacy by environmentalists. Many activists described being involved in the decade-long effort to stop logging in the North Shore watersheds and explained that as a result of activism on this issue, the Metro Vancouver Board created a permanent public representative position on the Metro Vancouver Water Committee (an advisory committee made up of elected officials, water experts, and community members) to allow input from environmentalists and concerned citizens.

Jennifer Brown, one of the leaders of the coalition against water privatization, is an environmental activist and chair of the organization Conservation Now, a local conservation group. She worked for over ten years to establish a ban on logging in the region's watersheds, to protect drinking water quality. As part of her advocacy work, Brown was instrumental in influencing the Metro Vancouver Board's decision to create a position on the water committee for a citizen representative, and she was the first person appointed to the position. She explained the importance of the relationships between movement organizations and government officials who oversee the delivery of water services to the region:

Winning the privatization fight had a lot to do with the fact that we had had our previous wins, and we knew the players and they knew us and when they saw us coming they realized they had to sit up and take notice; otherwise it would become a controversial issue that might lose them votes. . . . It was a really short battle. And it helped us get people involved too, because they knew who we were and trusted us and so were motivated to come out to the meetings. When [the Metro Vancouver board] saw the public coming they just changed. They really backed off.

Previous movements had created the networks and a mobilized public needed to respond to opportunities as well as the established alliances with elites that facilitated the openings for movement success.[22]

Beyond advocacy training, the presence of key allies within political institutional settings was critical for the movement's ability to have their claims considered within the decision-making arena. "People who were leaders in the movement were very connected politically. This wasn't an outside movement. It had a lot of connections on the inside. It was building the public opposition and then having friends inside to do the inside work, using the public opposition. That was a big part of it," explained Mike O'Brian, a national water campaigner for the Citizens Action League (CAL), one of the main organizations of the anti–water privatization coalition.

These preexisting ties between activists and political leaders were built upon, as the campaign also gained legitimacy by strategically targeting political elites who were not previously considered allies. Frank Dooley, one of the leaders of the anti–water privatization movement, explained that the members of the coalition thought it was important to seek support from city councilors who did not traditionally support the causes of the "left-leaning" organizations involved in the movement: "The support from councilors from Vancouver ended up being really important," he explained, adding that despite the pro-business stance of the Vancouver City Council, they were seen as "more attentive to community and environmental concerns" as well as "the power brokers of the region."

During the campaign to stop privatization, Frank Dooley worked closely with CAL, a national social justice organization with local groups across the region. Because of his involvement with a local chapter of CAL, he felt he was not in a position to convince more conservative politicians to consider the movement's claims; he worried that his efforts would be dismissed. He chronicled how some of the leaders of the anti–water privatization coalition decided it would be more strategic to solicit the support of someone who was considered an outsider to the social movement organizations that made up the coalition. Several of the leaders, including Dooley, approached Penny Blythe, an independent researcher who was seen as being "an objective outsider." The coalition members felt that this strategy would be more effective in getting conservative politicians to listen to the concerns about water privatization and local control of resources, particularly in the context of international trade agreements, such as NAFTA. He described the movement's efforts to work with those generally considered to oppose the issues supported by the Left, including Hugh Thompson:

Hugh Thompson was a really interesting character. In some ways he was probably the most right-wing pro-business councilor on Vancouver City Council, but he was his own man. No one could tell Hugh what to think. . . . We knew that if we could get him to read this stuff, that he would see it. And he read it, and he said that "this is not good, this is a mistake." And the thing about Hugh was that if you could convince him, everyone else on council fell in behind him. It was like Nixon recognizing China. I mean here is a guy whose business credentials are impeccable, his judgment unquestionable, and who is the hardest working pro-business councilor on council, and so if he says this is bad, it must be bad.

Targeting pro-business councilors, especially key political leaders from Vancouver, was pivotal on the part of the Vancouver movement in moderating the anti–water privatization claims and strengthening the movement's leverage within the institutional structures of Metro Vancouver.

Many activists in Vancouver also described the decentralized structure of Metro Vancouver as an important factor in creating openings for the movement to voice their concerns. Despite the pro-privatization stance of some of the Metro Vancouver Board members, respondents described the political culture in the region as flexible because of the dispersed nature of the decision-making power. Because the voting members of the Metro Vancouver Board are elected representatives from municipalities across the region and depend on the support of their local constituents for reelection, they tend to represent the interests of these constituents over the political views of the individual members of the board. Many respondents pointed to the decentralized representative structure of the Metro Vancouver Board for facilitating a more neutral position toward water privatization on the part of many of the directors.

The decentralized structure of Metro Vancouver was a particularly salient factor in the face of widespread and organized opposition from the public. The mobilization efforts, combined with the structural openings at the institutional level, were critical for allowing the movement to influence decision making. Peter Clark, one of the main leaders of the anti-water privatization movement, told me that the Metro Vancouver representatives were more vulnerable to mobilization because they understood that their political power lies with their local electoral base rather than through their position as directors of the regional district. A past employee of Metro Vancouver and former president of City Green (a local environmental organization), Clark described how mobilizing public pressure targeted at individual councilors serving as representatives of Metro Vancouver was important for allowing public input on water privatization. He described the passionate intensity of the crowds of people who attended the public consultation sessions:

I just remember that meeting with over six hundred people. And it was actually more because there were people outside who couldn't even get into the hall. [The chair of the water board] looked terrified. People were so angry. This was not an information meeting. This was a "We are going to lynch you if you do this" kind of meeting, and so that was what was interesting. I mean I've been involved in lots of issues where you have an information meeting and people come out and they are very polite, very Canadian, and they go home and think about it. Well, let me tell you, they had already thought about it. This was, "Don't even think about going there."

Targeting municipal councilors through organized mobilization forced local politicians to pay attention to the voice of their constituents and weakened their initial commitment to support the Metro Vancouver Board's proposal to privatize water services.

Many Vancouver activists echoed Peter Clark's sentiment about the importance of widespread mobilization and organized opposition. A forty-five-year-old mother of two teenagers, Jane Poole, who was an environmental activist and locally elected official at the time of the proposed privatization plans, attended the public consultation sessions organized by the Metro Vancouver Board in the spring of 2001. She articulated her serious concerns about outsourcing water to a private company, fearing the negative consequences on the environment and on local decision-making power to regulate resources under privatization. She explained that ever since she was a teenager, she had been active in advocating for policy change at the municipal and regional levels and understood the importance of public input into policy making. After years working with nongovernmental organizations, Poole finally decided to run for local office in her community, and she was twice elected as city councilor and served one term on the Metro Vancouver Board.

Jane Poole felt that the intense public pressure from local constituents was critical to swaying the opinion of members of the Metro Vancouver Board in favor of the movement opposed to water privatization. She explained: "Metro Vancouver is more antsy about political pressure than other organizations. When you get hundreds of people showing up at a meeting that translates into thousands of people who are pissed off about this, and you say to yourself why should I take the heat as a result of it?" Poole felt that Metro Vancouver representatives were swayed by public pressure because they needed the support of their constituents to be reelected. Preexisting alliances with political elites, the decentralized structure of Metro Vancouver, mobilization efforts on the part of the movement, and the presence of hundreds of protesters at the public consultation sessions were all crucial elements in opening up opportunities for movement success.

Interplay between International and Domestic Opportunities
In Vancouver, activists located their critique of water privatization in the
dominance of multinational corporations and the power of global financial regulatory institutions, including NAFTA. It is clear the movement
seized local opportunities, but activists also took advantage of opportunities and targets external to the domestic political arena—including
multinational corporations and international trade agreements—to open
up new opportunities at the local level. Many people described their concerns about the negative track record of international water companies
in other parts of the world and the profit-driven motive that has resulted
in increased rates for water services and a lack of investment in critical
infrastructure upgrades and maintenance.

Mark Spencer described how Metro Vancouver was under pressure
from multinational water companies to outsource water services. He
told me that in the years preceding the decision to privatize the water
treatment plant, several Metro Vancouver Board members had been approached by multinational water corporations at the annual meeting of
the Canadian Federation of Municipalities. "Executives from some of the
big players in water would be there to meet us and sell us on the idea of
P3s," he told me, adding that, "there were always councilors ready to buy
into the scheme." Spencer explained the role of multinationals in influencing local decisions about infrastructure and service delivery:

The pressure, in essence, comes largely from major multinational companies who
come in and want to propose these kinds of schemes. So they started to make
their approaches and usually the way they do it is they drag over someone from
England. You know, being Canadian we are always impressed with someone with
an English accent. It could be the dumbest guy but if you've got an English accent,
we think you are smart. And it's got to be a certain kind of English accent. Can't
go with Cockney because we've all seen *My Fair Lady*. It's got to be a BBC English
accent. Plus there are a lot of people out there who recognize that this was a place
that they could develop business, so there were a lot of consultant engineers and
people like that who suddenly jumped on the bandwagon about privatization.

He described how pressure from multinational water companies was instrumental in shaping local decisions about how to treat and deliver water.
Leaders in the anti–water privatization movement realized the increasing
power of global water corporations and their influence in local politics
and seized on those opportunities to target corporate power structures.

Activists articulated their concerns not just about the track record
and profit-driven agenda of multinational corporations, but also in the
mistrust of local governments and their capacity to regulate and control resources. Sean Becker, the thirty-nine-year-old community activist

introduced in chapter 4, who heads an organization that connects local issues with global concerns around democracy, human rights, and the environment, explained his lack of faith in municipal governments. "At a core level, it was really a mistrust and unease that those who are in the position of making decisions around this were way out of their league in terms of who they were playing with. Some of them I think get excited that they are playing in the big leagues," he said. Becker went on to describe the role of global firms and the concern some people felt with the government's ability to deal with multinational companies: "It's their job to squeeze public players and make us feel good about the fact that we're getting something out of it. And so there was definitely mistrust that those who were mandated to maintain the resources actually really know what they were talking about." Drawing attention to the risks of privatization to local political autonomy reflected a general sense of mistrust on the part of anti–water privatization activists in Vancouver about the ability of local governments to regulate corporate power once a public service has been outsourced.

Activists also targeted economic institutions by focusing on the role of corporations and the consequences of for-profit water delivery. Amanda Jones, a campaigner for Citizens Action League, explained how the movement targeted multinational water corporations in an effort to mobilize public opinion and prevent privatization. The movement focused on the negative track record of actions by Bechtel in particular, one of the multinational water companies short-listed for the private contract in Vancouver. Jones described how a subsidiary of Bechtel had been awarded the contract for water services in Cochabamba, Bolivia, where dozens of protestors had been injured and one person killed in a mass uprising against the water company and its decision to raise rates and cut off those who could not afford to pay for water:

Well, they had short-listed a number of large companies that were involved in international water scandals, and Bechtel was one of them and the most likely winner of the competition. And I think people really took issue with that and hammered them on it. . . . We said, "The reason that you are looking at contracting these companies is because of their global expertise. Well, what is the other side of that? It is corruption—it's all of those charges. These companies are involved in killing people who oppose their rainwater being privatized, and you really think that is the best company to run such a precious piece of the public commons?" So we used those arguments to smash their claims that these companies would be good for the local economy and good corporate citizens.

The movement targeted corporate power and seized opportunities from global examples of for-profit water treatment and delivery to counter the pro-privatization arguments of some of the Metro Vancouver Board members and staff.

Many activists in Vancouver discussed the importance of targeting corporations as well as politicians to demonstrate how multinational companies are implicated in decisions that affect local control of resources. Ronald Hudson is an environmental and social justice activist who works with local environmental and community organizations to fight the encroachment of the local economy by multinational corporations. He joined the fight against water privatization because he did not want to see a global firm control water in his community. Hudson explained that focusing on corporate power was important for the anti–water privatization movement, both for neutralizing the arguments of elites and mobilizing the broader public to take a stand against privatization:

I think that it's a weapon you can use against your targets saying, "You can't point to every place else in the world and say you were universally accepted with open arms," as some of these corporations try to do. They say, "Every community wants us in, we bring jobs," and they talk about what wonderful corporate citizens they are. Well, people know differently, and it is good to point out that "Well, why did these people over in Cochabamba kick you out then? Why did you get in that big fight in Atlanta where they said no to you?" If you are so great and you bring such benefits to the community, why are you being kicked out of communities?

Hudson pointed to the importance of targeting corporate structures and using the narrative opportunities provided by the experiences of other communities with multinational water companies for countering the pro-privatization arguments of local authorities and mobilizing people to join the movement.

While anti–water privatization activists articulated their concerns about the dominance and power of multinational corporations, they also targeted international financial and regulatory institutions that govern trade in goods and services between countries and protect foreign investment. Movement actors were concerned that the regulations under NAFTA would prevent the regional government from regulating water resources to protect a foreign company that operated the water system. "We focused a lot on the implications under NAFTA. And it was difficult at first, particularly with politicians, to get them to understand there were all these implications from trade agreements," said Sandra Gibson, a forty-eight-year-old office manager for the regional office of the Citizens Action League. She described how the movement commissioned a legal opinion on the ramifications of privatizing water under NAFTA to challenge the arguments of Metro Vancouver officials and convince them to support their cause: "And part of it was that the government could be prevented from regulating water or protecting water if it meant a loss of profit for the company involved. So once we went down that road, it would be a slippery slope in terms of maintaining local control of our water." Providing legal evidence of the risks to local control under NAFTA captured the attention of local elites and created openings for the movement's claims to be considered.

Seizing global opportunities and expanding targets beyond political institutions to include the broader political economy—including economic institutions such as corporations and international financial and trade agreements—and bringing those concerns down to the local level, also influenced the tactics used by the movement in Vancouver. Activists drew on the tactical repertoires of broader anti-globalization, anti-trade, and human rights movements, citing international law and using disruptive, creative, and highly visible tactics borrowed from the anti-globalization protests.

Sherry Carruthers described the importance of global connections and the use of legal and disruptive tactics:

I think that the international connection to corporate greed and trade agreements made the movement what it was. The meetings were packed. And they were loud. That clearly galvanized people. . . . Water privatization was seen as part of that broader globalization trend, so I think that is why it caught such a fire and why people were just so pumped up. . . . We made that specific case about the potential liabilities under international trade agreements and why this wasn't a wise route to go. . . . And in the end that is what resonated with Metro Vancouver folks.

By connecting the movement to broader issues of globalization and international trade, activists in Vancouver not only mobilized a wider pool of movement supporters, but also succeeded in swaying the opinion of the Metro Vancouver Board.

The combination of legal and disruptive tactics helped shift the opinion of Metro Vancouver representatives against privatization. Richard Martin, a senior Metro Vancouver bureaucrat, explained that the emphasis on threats posed by multinational corporate policies and international trade agreements, including the impact of NAFTA on local political power, as well as the range of disruptive tactics were pivotal to the decision by the region's water board to reverse plans to privatize water services in the region. Despite this reversal, he described his surprise at the enormous turnout at the public information sessions and the intensity of the opposition to water privatization. "The meeting was scheduled to start at 8:00

and there were only about fifteen people there at five minutes to 8:00. We thought this was going to be a total breeze," he said, describing the first public information session in Burnaby. Shortly before the meeting began, just as he was about to begin his presentation, Martin began to hear drums beating. "And then literally hundreds of people came into the room . . . and they were noisy, and there were drums and there were people dressed up as ninjas and fancy costumes and goodness knows what. All hell broke loose. There were people dancing around . . . and people would run up and steal the mic off the guy who was trying to facilitate it," he recalled. In fact, there were so many people that the security staff began blocking people from entering the room for fear of breaking fire regulations. Martin described the evening as a "pretty hideous experience" because of the unruly and disruptive behavior by anti–water privatization activists.

Despite his disagreement with the choice of tactics, Richard Martin said that he "started to think twice about the direction we were headed" because of the concerns about the legal ramifications under NAFTA that activists presented at the meeting. "So although we weren't entirely convinced, we were concerned enough and also knew we were facing such enormous opposition, that we decided it probably was not worth the political storm in the end," he explained.

The seizing of international opportunities and targets by the anti-privatization movement in Vancouver shaped tactical repertoires that worked to highlight the vulnerability of municipal governments in the face of the growing power of global economic institutions. The attack on corporate power and targeting of opportunities beyond domestic political structures by the movement in Vancouver reflects the growing awareness on the part of activists of the multiple sites of institutional power that shape movement opportunities and constraints, and adds a new dimension to the political opportunity structure model that moves beyond political structures to include corporate targets and international economic institutions.[23]

In Vancouver, the relative openness of the political institutions involved in the decision making around water treatment and delivery, and the established activist-elite alliances created divisions within the local polity. The seizing of global opportunities and the use of disruptive tactics facilitated broad mobilization and opened up opportunities for movement claims to be considered, and ultimately shifted the opinion of Metro Vancouver Board members in favor of keeping regional water services under public control.

Conclusion

Despite the hegemonic power of neoliberal globalization, how it resonates at the local level and how countermovements mobilize is not uniform across different contexts. In response to similar privatization threats, differences in the degree of institutional openness in Stockton and Vancouver led to diverse responses to the problem by the anti–water privatization movements in each place.

In Stockton, anti–water privatization activists were constrained by the relatively closed nature of local political institutional structures, which prevented activists from creating alliances with political authorities. The municipal government—led by a conservative mayor—remained unwilling to consider movement claims and ultimately voted in favor of privatization, without any public input. Growing pressure from the anti–water privatization movement forced them to rush through the vote on privatization. Beyond institutional closure, the unfavorable conditions in Stockton were exacerbated by the lack of preexisting movements targeting the local polity and the relative inexperience of activists organizing at the local level. A narrow focus on legal and nondisruptive tactics prevented the movement from seizing international opportunities, including anticorporate and antiglobalization arguments, and blocked efforts by some activists to mobilize a broad-based grassroots opposition movement. Without widespread mobilization or the integration of international concerns into the repertoire of contention, the movement in Stockton was unable to mobilize the kind of public opposition needed to influence policy decisions, or create a sense of local solidarity in the face of global threats. Yet, through a legal challenge, the movement was eventually able to overturn privatization, creating a political opportunity in a closed system, and achieve success. Their eventual success points to the importance of studying movement outcomes in the longer run as well as the need to adopt a broader understanding of movement outcomes, beyond evaluating the success or failure of a movement's initial goals.

The anti–water privatization movement in Vancouver, in contrast, faced more favorable opportunities for the consideration of grievances because of the relatively high degree of institutional openness of the regional government, divisions among elites, and alliances with authorities. The success of the movement in Vancouver was facilitated by more than structural openness. Preexisting movements, including the watershed protection and anti-globalization movements, created alliances among activists and local elites and shaped the choice of targets and tactics utilized by the anti–water privatization movement. Activists seized global opportunities, including multinational corporate policies and threats from international trade agreements to increase mobilization and create a sense of local solidarity in the face of global concerns. Faced with widespread opposition, authorities in Metro Vancouver agreed to provide a forum for public deliberation, opening up critical opportunities for movement success.[24]

Political opportunities are available to movement actors at different institutional levels, including sites of local and global power. International opportunities strengthen local movements by fostering ties between domestic actors and creating a sense of solidarity among local authorities and activists. Anti–water privatization movements provide a useful case study for understanding multiple opportunity scales because they are locally rooted and dependent on domestic institutions, while at the same time are implicated in multinational economic power structures that influence the nature and availability of political opportunities on the ground.

Activists at the local level can draw upon global threats and issues and apply them in ways that make sense in their communities, creating a sense of local empowerment that strengthens the capacity of local social movements to organize, mobilize, and respond to local opportunity structures. Movements that seize global opportunities and synthesize them with local opportunities, as in the case of the anti–water privatization movement in Vancouver, create a sense of *local solidarity* that allows communities to overcome divisions between authorities and activists, who unite in a common desire to protect local resources from threats from multinational corporate power and international trade provisions that protect foreign investment.

Recent research on social movements suggests that in the face of economic globalization and the growing power of capital, movements are increasingly seizing opportunities provided by economic structures and targeting institutions beyond the political realm.[25] Some movements recognize the power of transnational economic institutions for shaping domestic policy and shift targets accordingly; at the same time, many activists remain committed to reaffirming the role of the state in regulating social and environmental policies. In Vancouver, although the anti–water privatization movement seized opportunities provided by international economic structures, including multinational water corporations and international trade agreements, they drew on these opportunities only insofar as they would enable them to gain the attention of local authorities; throughout the campaign, the target remained squarely fixed on local political structures. This focus on the local polity can be explained by the fact that water was the central focus of the movement. As water is considered a public resource that is part of the commons, and the goal of anti–water privatization movements is to *prevent* commodification, the main target of such movements is likely to be the political institutions that govern and regulate water resources and services, rather than multinational water corporations and financial institutions regulating trade and investment.

Movements should choose targets according to the nature of the "product" that is the focus of their campaigns and the evidence from the movement in Vancouver reveals a need to refine the economic opportunity structure model to include differences between the commons and commodities. In the case of environmental resources, particularly those that are part of the commons, such as water, the polity has the power to protect and regulate these resources; hence, it is more likely to be the target of anti-privatization movements. Economic structures are more likely to be used as *symbolic targets* to highlight the political vulnerability of governments. To anti–water privatization activists the commodification of resources is not seen as inevitable. What matters to movements focused on protecting the commons is reasserting the power of the state to uphold democratic and public control of these resources and protect the sovereignty of communities in light of economic globalization. While activists in both Stockton and Vancouver were critical of political authorities and the push to privatize water, they recognized the power of government to protect environmental resources from the encroachment of private capital.

The traditional assumption is that the nation-state is the site of power, both instrumental and coercive, with the strongest capacity to ensure the well-being of its citizens. But in the face of economic globalization, it is local governments that potentially have increased leverage over national governments because they are not signatories to global governance agreements and thus are less constrained by the power of international economic institutions. Local governments have the capacity to retain an important form of democratic power in the face of pressures from international financial and regulatory bodies, especially in the context of environmental resources that are embedded in local economies and geographies.

As the power of nation-states declines in the face of a growing complex global political-economic order, cities have become key sites of political, social, and economic power; they are the engines that spur growth, attract investment, and draw new immigrants.[26] Cities are crucial for addressing the complex problems facing humanity, including the climate crisis, food security, and growing poverty and inequality. For example, Harriet Bulkeley and Michele Betsill argue that cities are essential for shaping and implementing climate change policies. While the focus of climate change policies over the last twenty years has been predominately at the international policy level, Bulkeley and Betsill maintain that cities—as important producers and regulators of greenhouse gas emissions—are well positioned to mitigate and adapt to climate change through adopting urban sustainability measures. They claim that local governments should play a central role in developing strategies to combat climate change because they "exercise a degree of influence over emissions of greenhouse gases in ways which impact directly on the ability of national governments to achieve internationally agreed targets."[27]

Global cities also generate new urban social movements, made up of activists who are rooted locally and committed to the health and well-being of their communities. At the same time, they are embedded in national and international institutions and enjoy a sophisticated understanding of local-global interconnections. Pierre Hamel, Henri Lustiger-Thaler, and Margit Mayer argue that these local movements reflect a new political culture and have the capacity to generate new modes of decentralized governance based on public participation and negotiation between different levels of decision making.[28] Local social movements have greater access to elites and operate in more participatory forms of deliberation at the local level. This means establishing concrete targets and clear channels for movements through which they can influence policy and strengthen local political power to stand up to challenges from economic globalization. The anti–water privatization activists in Vancouver responded to this challenge, demonstrating that social movement actors are aware of this power dynamic. The opportunities are there to be taken; the result will be a stronger democratic process.

6

Mobilizing Cross-Movement Coalitions

Is there a connection between the labor movement and the environmental movement? I haven't seen it yet. Again, the labor movement is going to be very pragmatic in how it looks at things, and not to denigrate the environmental movement, but my view of the environmental movement is that they still have that kind of iffy-based mentality about minimum wage types of issues, and those kinds of things where unions have labored hard. We were born out of, seriously, people losing their lives and those kinds of things. So our view of it is a hundred years ahead of the environmental view.

—Ken Bernardo, union representative, Stockton

I think it is starting to come. I think environmental groups know that things like climate change are starting to get taken more seriously, and are becoming more cognizant of policies and how they play out on the ground, and so then are more concerned about the social justice side of things. And I think a lot of trade unions are becoming greener in terms of how they think about things. We need to bring together unions and the resource sector with environmentalists and First Nations . . . to talk about common ground and how we can have a sustainable resource sector and good, well-paying jobs. There has been a lot of groundwork that has been laid but they are still coming together. To me the politics of the future is around those two elements coming together in a concerted way.

—David Smith, economist and environmentalist, Vancouver

In many communities around the world—from Cochabama, Bolivia, to Orange, South Africa—anti–water privatization movements have resulted in the formation of coalitions between labor unions and other community-based social movement organizations.[1] These coalitions vary in strength and outcomes, and opposition movements are less likely to succeed when they have not formed these broad coalitions. Cross-movement coalitions, including the involvement of labor and environmental organizations, emerged in both Stockton and Vancouver in response to proposed water privatization plans in each city. The coalition in Stockton never gained the strength, momentum, or positive outcome of the Vancouver coalition, and quickly abated after a promising start. Why? Movement participants suggest it was the critical role played by labor union culture and frame-bridging organizations. Perspectives of two key labor leaders from each location reveal important clues for understanding the divergent trajectories of the coalitions in Stockton compared to Vancouver.

Stockton

When the city of Stockton announced plans to contract out the operations and maintenance of its municipal water services in 2001, the coalition that came together to oppose the outsourcing included representatives from diverse social movement sectors. These included environmental organizations, civil rights and voter advocacy groups, and labor unions. During the initial stages of planning, the coalition reached out to employees of the municipal water utility in an attempt to engage them in the movement and help voice their concerns over privatization. One of the employees who heeded the call was Vince O'Neil, a fifty-three-year-old senior plant maintenance worker. O'Neil joined the coalition's efforts because he was tired of the portrayal of public employees by the mayor and city council as "fat, lazy, dumb workers who come to work every day and lean on a shovel." He believes government employees should have a say in political decisions about the running of public utilities, and became involved with the coalition against water privatization in Stockton primarily to ensure that "from the employee point of view, fair and balanced information was getting out to the public."

Born and raised in Stockton, Vince O'Neil is a grandfather and the full-time caregiver of his two-year-old granddaughter. He has been maintaining the pumps at the Stockton water treatment plant for more than thirty-two years, and is a member of the Utility Workers Local 5, the union representing municipal utility employees in Stockton, as well as president of his union's association. Despite having voted Republican most of his life and believing in limited government, he has faith in the value of the public sector and takes great pride in the work that he and his fellow employees do at the plant. He often puts in extensive overtime hours, returning to the plant in the middle of the night to complete critical repairs when no one else is available, although he does not single himself out as unique in this regard, saying, "The bottom line is there are a lot of employees that do what I do. Nobody talks about it, and we are not out there to get any accolades for it." As an active member of his union local, including having held an executive position for many years, O'Neil wants to see municipal utility workers treated fairly, since they are "gifted employees who come to work every day dedicated to doing a good job and willing to put in time after work." He credits the municipal utilities department for turning his life around from his self-described "hellion" days when he "drank a lot, drove a motorcycle, and was crazy," and describes the department as the place where he grew up "from a boy to a man," among dedicated people who had a strong emotional attachment to the job. At the same time, he worries about the potential health and safety risks of working at "a dirty job, a filthy job, on the wastewater side of it."

Vince O'Neil was not previously acquainted with the leaders of the citizen's coalition that formed in opposition to water privatization. Although he initially worked closely with coalition members outside of the labor movement and has great respect for the work they do, his focus soon shifted to fighting the private company that took over his workplace so that he could help protect plant employees under the private contract. While he is concerned about environmental issues, particularly about multinational water companies' "dismal track record on the environment," he is not a member of any environmental organizations and does not identify with the environmental movement. In fact, he felt frustrated with environmentalists for not recognizing what matters most to workers' lives, such as on-the-job health and safety, decent wages, and job protection. Consequently, his involvement soon waned because he felt that workers' issues were dismissed by the coalition.

Vancouver

In the fall of 1999, when rumors first surfaced that Metro Vancouver was planning to privatize the management of water treatment and delivery systems, the British Columbia Public Sector Employees Union (BCPSEU)—one of the largest public sector unions in Canada—began to organize a public opposition movement. Leading the campaign was Heather Harrison, a lawyer by training and long-time community activist in the Vancouver region, who at the time of the proposed water privatization plan, had been working as a researcher for the BCPSEU, specifically focusing on the privatization of public services. Harrison, a fifty-five-year-old mother of five adult children, works full time, including putting in considerable overtime hours at the BCPSEU office, which occupies three floors of a building in suburban Vancouver. The office houses a large research library with many resources on the importance of public services and the dangers of privatization.

When I interviewed Heather Harrison at the union headquarters, the office was buzzing with talk about the current government push for out-sourcing public sector jobs. Posters highlighting a range of issues—from the link between public ownership and democracy to the importance of public investment for environmental regulation—lined the walls of the hallway. It was clear that a significant amount of the union's resources are dedicated to issues beyond the work-based concerns of its members, including those that matter in the broader public policy arena.

A Freedom of Information request in 2001 revealed that Metro Van-couver had been exploring water privatization since 1995, and was aware that there would be widespread public concern about the issue. Harrison sprang into action, knowing there was little time to waste in getting the union's message out to the public before privatization plans were final-ized. She began to focus her research exclusively on the dangers of mu-nicipal water privatization. The issue was considered so important to the BCPSEU that she was provided with the full-time support of two other colleagues—a fellow researcher and a communications staff member. The union president also devoted a significant amount of time to the issue, speaking at public meetings and reaching out to the media.

Heather Harrison strongly believes that political change will only occur as a result of individuals and organizations uniting to form a strong oppo-sition to what she calls the "corporatocracy" that she feels "is pushing for trade agreements that will increasingly take away room for government to make public policy decisions in the best interests of the people." As part of the BCPSEU mandate, she regularly works with community groups and social movement organizations outside of the labor movement on issues that are considered of significant importance to the wider commu-nity, including the environmental, economic, and social impacts of neolib-eral globalization. She described the anti–water privatization campaign as "a load of fun and very empowering," and saw it as a natural extension of her previous work as a community activist. Her involvement in the politi-cal community—her husband is active in local politics—and strong ties to environmental and social justice groups throughout the region gave her a strong network base from which to mobilize. The BCPSEU also has a history of reaching out to social movement organizations to create "solidarity with the community," working for over a decade on fighting the privatization of water services as part of Protect Public Water!, which brings together labor, environmental, and social justice groups, describing it as "a very natural coalition for us to be involved with."

Heather Harrison and her colleagues at BCPSEU recognized the im-portance of building a broad-based coalition and immediately began to solicit community support. They commissioned a poll showing widespread public opposition to privatization, held high-profile community meetings with environmental leaders, and organized information packages about water privatization and the risks to public control that were sent to several prominent community leaders and local politicians. "Community coalitions were huge. We knew how important this was so we got a list of all the community groups in Vancouver. . . . There were seniors groups, neighborhood groups, youth groups, some churches—all sorts of community groups. And we sent out packages of materials explaining what our concerns were to around seventy community groups," Harrison explained. These organizations, many of which have already been introduced in previous chapters, included City Green, a local environmental organization that was heavily involved in watershed protection issues; Conservation Now, a prominent conservation group that had worked closely with the Metro Vancouver Board; and the Citizens Action League (CAL), a large citizen-based social justice organization.

According to Harrison, the BCPSEU was able to attract significant public and media attention because they focused on issues of "public control, privatization, and globalization, and water quality and service," a strategy that was "really effective in mobilizing public support." What was also significant, she explained, was that the BCPSEU engaged in the anti–water privatization movement even though they did *not* represent the public sector Metro Vancouver employees at the water treatment plant, whose jobs would be affected by privatization. This strategy was effective because it increased the union's credibility with both the public and political elites who "couldn't argue that we were doing it to protect jobs and workers." The BCPSEU was also strategic about their role, providing more resources than visible leadership in the coalition. Throughout the campaign, Harrison and her colleagues were conscious of maintaining a low profile in terms of leadership so that their efforts would not be dismissed as "just another union campaign."

By reaching out to the wider community, the campaign "took off, it just resonated with people," Harrison said. Those who attended the public consultation meetings were representative of a diverse group of citizens. She described the meetings as "astonishing" in terms of the sheer numbers of people and whom they represented. In one case, the fire marshal was forced to turn dozens of people away because of overcrowding. Harrison told me, "There were so many different groups and people. There were young people dressed with blue paint on their faces, but it wasn't just young people. There were a lot of seniors . . . and there were also just a lot of people that were concerned that just showed up that didn't want their water privatized." For Harrison, the widespread public representation at the public hearings signaled that the movement had moved beyond a union battle to become a true community-based opposition movement.

Cross-Movement Coalition Building in Social Movements

The anti–water privatization movements in both Stockton and Vancouver included coalitions of community organizations, such as labor and environmental groups, that had frequently been on opposing sides of many issues in the past. Although coalition building was a major factor in the resistance movements in both cities, as we have seen, the trajectories of the community alliances took on diverse forms and outcomes, with the movement in Vancouver reflecting a stronger, more unified movement than in Stockton. Specifically, labor was an integral part of the anti–water privatization coalition in Vancouver; in Stockton it eventually withdrew. This divergence in union participation provides important insights into the dynamics of coalition building—particularly labor-environmental coalitions—in social movements.

Labor-environmental alliances are more successful when the labor unions involved are guided by a social movement unionism model, as in the case of the Vancouver movement, versus a traditional business unionism model.[2] Because of their capacity to unite previously disconnected groups and synthesize frames, in the context of localized resistance to neoliberal globalization, the presence of key bridge-building organizations—social justice groups in particular—is essential for successfully building labor-environmental coalitions.

Coalition Building in Response to Public Sector Restructuring

The rise of the neoliberal agenda globally, and the push for private sector involvement in areas once considered the domain of the public sector—from education to water treatment and delivery—have generated new threats to public sector service delivery and the management of natural resources. Community coalitions that form in opposition to the privatization of public services—including water—are often made up of diverse social movement organizations and actors, including representatives from labor, environmental, and social justice movements. Increasingly, these organizations and movements have created coalitions around campaigns that oppose privatization. Recently, many scholars have argued that local, broad-based community coalitions—especially those that unite labor and environmental movements—represent the countermovement that is needed to oppose the destructive social, economic, and environmental policies of neoliberal globalization and promote social and ecological justice.[3]

A critical example of private sector involvement in public services is the outsourcing of water treatment and delivery in municipalities across the United States and Canada and globally, an area that has been largely the exclusive domain of the public sector.[4] While the trend toward restructuring municipal water services threatens public sector workers in terms of both wages and job protection, it also creates the potential for coalition building between unions and other social movement sectors, including environmental and social justice movements. Unions and other community organizations frequently come together to fight what they perceive as a threat to local control of resources and services, especially in key areas such as health and the environment.[5] Water privatization as an issue resonates across multiple dimensions, encompassing economic, social, and environmental concerns, and thus offers the possibility for previously disconnected movements and organizations, even those that once positioned themselves on opposite sides of issues, to work strategically together to protect public water resources and services.[6] Water service restructuring also offers a clear and tangible target for grievances, and coalitions are more likely to develop with the presence of easily identifiable opportunities or threats.[7]

The environmental movement should be a key target for alliance building with organized labor because of its widespread appeal and—similar to the labor movement—its high degree of power and influence in the decision-making arena relative to other social movements.[8] Despite the very real divisions and adversarial relations that have historically occurred between the two movements—generally perceived as the "jobs versus the environment" problem[9]—there are new opportunities for labor and environmental movements to work together on certain issues or campaigns in the face of growing threats from common foes, including advocates of neoliberal policies and reforms. Workers are facing increasing threats of job loss, outsourcing of public sector jobs, and downsizing, while multinational corporate policies and obligations under international trade agreements threaten the ability of local governments to enact legislation to protect both workers and the environment. As more and more labor unions adopt social movement unionism over traditional workplace organizing and environmental movements increasingly recognize the importance of environmental and social justice concerns, the links between these two movement sectors continue to strengthen.[10]

Some scholars argue that while unions have historically been resistant to supporting environmental policies over fears of job losses, a new and revitalized labor movement that embraces cross-sectoral coalitions is key to reversing the ecological degradation caused by the continuous expansion of the global capitalist economy.[11] Brian Obach contends that the new labor movement has the potential to build relationships and common ground with the environmental movement in order to advance economic and social justice. Recent transformations, including the shift from manufacturing to a service-sector economy, the expansion of neoliberal globalization, and the emergence of a younger leadership that is open to building alliances with other movements, have fostered new ties between labor and environment, and created possibilities for innovative economic policies that do not harm the environment. At the same time, Obach cautions that the environmental movement must also work to build bridges with organized labor by incorporating environmental and social justice issues into their campaigns and advocacy work.[12]

The environmental movement has not traditionally been supportive of the concerns of labor unions or working class communities. Several scholars, including Mark Dowie and Robert Gottlieb, have argued that the environmental movement in the United States—while highly visible and enjoying widespread support—has been largely ineffective during the last three decades at halting or reversing ecological degradation. Dowie contends that this is due to the less radical, more conciliatory approach of the mainstream movement toward the Reagan and Bush administrations that resulted in a reversal of earlier successes and the unwillingness of the movement to embrace concerns of working class or minority populations.[13] At the same time, as Gottlieb argues in his book *Environmentalism Unbound: Exploring New Pathways for Change*, the predominance of large, professionally run environmental organizations, composed mostly of white, middle class people, has disenfranchised and alienated a large part of American society. Among the groups excluded are low income individuals, ethnic minority immigrants, and people of color, who face serious environmental justice challenges largely ignored by the mainstream movement, such as workplace safety, healthy communities, and food security.[14] Both Dowie and Gottlieb maintain that the next wave of environmentalism will prevail as long as it is more radical and inclusive than the previous wave, and bridges mainstream pollution and conservation concerns with issues of environmental and social justice.[15]

The contrasting stories at the beginning of the chapter illustrate the different views of labor leaders in Stockton and Vancouver toward environmental and community organizations. In Stockton, the Utility Workers Local 5 (UW5) union, with its focus on workplace concerns, was more detached from the other community organizations that participated in the coalition. In contrast, in Vancouver, the BCPSEU strategically decided to reach out to organizations outside of the labor movement by actively engaging these organizations and framing their argument on issues with broad social appeal instead of traditional workplace concerns. How do these coalitions work, and what makes some more successful than others? To answer these important questions, we must examine the processes that underlie their development, including how and why they emerge, how network density and cohesion facilitates mobilization, and what creates the conditions for successful outcomes.

Collaboration, Coalitions, and Networks: The Role of Labor
Part of the tension concerning labor-environmental coalitions often arises from the conflicting focus and goals and the individual organizational cultures involved in each movement. Historically, much of the antagonism between the labor and environmental movements has been rooted in class-based differences, with labor focusing on traditional working-class concerns—such as jobs, wages, and benefits—and the environmental movement mobilizing around issues that were often perceived as anti-union and a threat to jobs, including wilderness protection, conservation, and the regulation of toxic pollutants.[16] Over the last decade, there has been a waning of conflict (a détente of sorts) as cross-movement collaboration grows, and each movement, independently, is revitalizing. This shift reflects, in part, a change in organizational culture. Social movement unionism strengthens the political power of the labor movement by facilitating grassroots organizing, access to resources, and coalition building with the broader community.[17] At the same time, local, grassroots environmental organizations have the potential to reinvigorate the environmental movement by focusing on building coalitions with diverse organizations around local political issues.[18]

One of the major factors explaining the divergent trajectories of the Stockton and Vancouver movements is differences in the organizational culture of the unions and other organizations involved. In Stockton, UW5 was involved in the formation of the coalition against water privatization, but in concentrating attention and mobilizing efforts on workplace concerns, their role soon diminished as the movement solidified. The union was unable to connect its priorities and focus to the goals and frames of the other groups in the coalition. At the same time, nonunion organizations were reluctant to embrace the concerns of the labor movement. In Vancouver, on the other hand, the BCPSEU focused on issues beyond workplace concerns and thus were able to play a key role in the coalition by synthesizing their arguments and goals with that of the broader public.

Stockton

In July 1997, the mayor of Stockton visited the city's Municipal Utilities Department (MUD), which oversees wastewater treatment and drinking water delivery services, and, in front of the public sector employees and managers, announced plans to seek proposals from the private sector for the operation of the utility. He argued that as a public utility, the department was a drain on the city's budget, was inefficient in its operations, and would be better managed by the private sector. The mayor's visit was the "thrown gauntlet" that represented the beginning of the a long battle with city hall, according to Ken Bernardo, who at the time was a senior wastewater treatment plant operator and chief steward of Utility Workers Local 5, the largest construction trades local in the United States, representing workers from four states. Bernardo is a long-time union activist with years of experience working in the private sector. Based on this experience, he told me that he often agrees with private sector criticism of publicly run services and systems, claiming that there is some truth to the argument that the private sector is more efficient. While he had ongoing issues with the MUD management, and did not disagree that there were inefficiencies in plant operations, he felt that the employees were unfairly attacked and viewed the situation as "more of a management problem than [as] . . . an employee problem."

As chief union steward, Ken Bernardo became centrally involved in the fight against privatization. His main goal was to "negotiate on behalf of the welfare and livelihoods of the employees" and ensure "the fair treatment of workers in the face of a double attack: a battle with . . . management as well as with the mayor and the business community that was backing [him] into the privatization of the treatment facilities." When he was approached by a group of individuals from the wider community asking for the union's support in the fight against privatization, he agreed to work with them, but remained focused on ensuring a fair deal for the workers, whom he describes as being worried about their jobs and their future at the department: "They felt like the assault was on them and their livelihood, and they were concerned that the change and all of that was going to mean they would be attacked. I think like the other classic institutionalized employees, they don't like to hear the burden is on them. They were worried about their jobs, worried about their income, having to actually perform change. All of those things were on their minds." While he believes that union struggles have a positive role in creating "a healthy society and a fair society," he felt the unions were primarily involved in the anti–water privatization movement to fight for job security and to protect wages.

Ken Bernardo explained that although the unions were part of the community coalition, they played a less visible role than the other coalition partners because of the perceived threat to their jobs both before and after the plant was privatized and the concern that they would lose their jobs if their involvement in the movement became known. Although he argued that the employees were discouraged when their jobs were privatized because of their concern that wages and benefits would decrease under the new contract with the private company, he described the union's focus on negotiation and job protection as being successful because they were able to secure a higher hourly wage for workers with the private company than under government control. He explained that, in general, the employees were not worse off after privatization:

The pay was better in the private sector because we were able to negotiate. They wanted to throw money at people to quiet us, so they threw a lot of money at us. We lost some protections from termination and those kinds of things. But, on the other hand, things could be done quicker. I think there are a full group of employees out there who think that we are actually better off under the private sector because we get paid for any disagreement or anything like that. The city can't do that as a public entity. They can't make a gift to us of public funds, so cash settlements and those kinds of things were not really . . . in the cards.

Many of Ken Bernardo's colleagues who were involved in the anti–water privatization movement in Stockton shared his traditional view of the role of unions and the reasons for engaging in the movement to prevent water privatization. John Sandler, a fifty-three-year-old city employee, was a maintenance supervisor at the water treatment plant at the time of the proposed privatization. He was a high-profile union leader and joined the coalition against privatization because of his concerns that privatization would reduce job security and worsen work conditions of the plant workers. Sandler is an extrovert; he exudes confidence and considers himself a leader in the fight to block privatization. As he explained, the employees at the Stockton wastewater treatment plant "looked to me for direction and guidance because of my leadership abilities at work."

With nearly thirty years of work experience at the plant, he feels strongly connected to the natural environment and that it should be protected for future generations. At the same time, he does not feel connected to the environmental movement and is not a member of any environmental organization. Sandler's motivation for joining the coalition was deeply personal: "I was [at the plant] for twenty-six years. There was a lot of pride of ownership with all of the employees. It was theirs, this is ours. You can't put a price tag on that. It was a very tight-knit group . . . this is what we planned on doing until we all retired." He went on to explain:

There is something about wastewater . . . seeing what the product looks like coming in the door, and what it looks like going out the back end. And most of us have boats, and we fish in the delta. There is something about seeing it, and saying "I did that." . . . Ninety percent of us could have gone and worked elsewhere, but that is what we chose to do because we genuinely loved what we was doing.

Many of the workers expressed similar sentiments. John Sandler felt that his pride in delivering a public service would not translate into working for a private for-profit corporation.

Sandler explained that once the plant was privatized, the workers felt "worn down" and "no longer had their hearts in their work." Personally, he resented the mayor's disparaging remarks about the work ethic of the plant employees and resigned his position in protest before the private company took control. He said, "I couldn't get myself up every morning and look myself in the mirror and accept this. . . . I flat out refused to work for the private company because of that. I couldn't go in and make money for them, after they told me how inefficient and ineffective I was." A focus on work efficiency and defending the pride of workers were common themes expressed by many of the plant workers and union representatives interviewed.

Senior plant maintenance worker Vince O'Neil also expressed the desire to protect workers, and highlighted the pride of public service as his motivation for getting involved on behalf of the union. He gave the following reasons for joining the movement:

One, on a personal level, because I took a lot of pride in that department. . . . The other reason I got involved was because I am not one [who is] afraid to speak my mind. A lot of the guys came and asked me if I would put together the association from the union side to help fight for our jobs because we were going to disband from the rest of the city. . . . So from the union's side, as president of the association, I just wanted to see things done right for the workers. . . . We have a tremendous amount of gifted employees, who come to work every day dedicated to doing a good job.

Similar to many other union members interviewed, O'Neil described how the union focused on the risks to the proper functioning of the water treatment plant under a private corporation:

We knew the private sector doesn't come in and do anything to break even. They are there for profit. In order to make a profit, your biggest payroll issue when you are running a company 99 percent of the time is personnel, staff. It's your payroll, it's your benefits, it's all this. So we knew coming in that the first thing that was going to be cut would be staff. . . . And in order for them to make this profit what was going to suffer was maintenance. The infrastructure was going to suffer.

Vince O'Neil and John Sandler's responses represent major motivations behind the initial involvement of the labor movement in Stockton's anti–water privatization coalition: to protect their members and highlight the dedication of the workers and the ability of public sector workers to deliver services efficiently and effectively. The quality of work and plant-based concerns did not naturally align with the interests of the other organizations in the coalition that were the driving force, such as environmental and voter advocacy groups, which focused on issues of accountability and environmental risk.

In Stockton, the union clearly operated largely under the traditional business unionism model, with a focus on addressing grievances, securing contracts, and looking out for the best interests of the employees it represented. There was little analysis or understanding of the wider issues beyond job protection and union pride. For union representatives in Stockton, the anti–water privatization fight was about job protection. It was about worker morale. It was ideological—pitting management and elected officials against unionized public sector plant employees. Although the plant workers and union members that I interviewed participated in and expressed support for the grassroots community coalition and for public control of water, the prevailing culture of business unionism prevented them from moving beyond narrow workplace issues and interests.

As a result, when it became clear to the employees and their representatives that the city government would move forward with privatization, the union's initial support and involvement waned—in part because union members were discouraged and demoralized, but also because their attention turned to negotiating a strong collective agreement with the private company. The reliance on a business unionism model meant that the union representing the plant workers quickly shifted their focus away from fighting privatization and into a strategic bargaining position with the new private employer to secure the best possible deal for the remaining plant workers.

Community-Labor Coalitions: Collaboration and Tension
Although the connection between unions and others coalition members was relatively strong at the beginning of the anti–water privatization fight in Stockton—the unions provided financial resources, meeting space, and leadership on community organizing—the relationship eventually became strained. While the unions focused on protecting the jobs, wages, and benefits of its workers with the private company, the coalition continued to oppose and fight to overturn privatization with new tactics, including a legal challenge. After the withdrawal of the unions as a central partner in the anti–water privatization movement, the remaining groups became disheartened with the labor movement. Many of the coalition members believed that the unions were mainly working for their own self-interest, rather than for the benefit of the community as a whole. This perception was widespread among Stockton movement activists who weren't part of the unions. I asked Joan Davidson, a former elected representative, local small-business owner, and member of the coalition steering committee, if there was a strong connection between the labor movement and the other community groups involved in the coalition. She replied,

I don't see any direct or strong connection. I think this one was self-preservation. The unions wanted to stay within the city of Stockton for a variety of reasons; they did not want to be employees of a private company. So that was one of those alliances made out of necessity . . . but, in general, I don't know. You'd have to ask union leadership about it. But the [Utility Workers] union, they were working very hard to keep the private company out because obviously it was about jobs and workers here.

Joan Davidson is a tall, well-dressed, sixty-year-old businesswoman with a professional demeanor. She moved to Stockton from Los Angeles thirty years ago to raise her family and is active in the business community and in local politics, running her own small business and having served as an elected representative. I interviewed her in a small shop she owns, which is located in a rundown, depressed area of the city. Just off the freeway, the store is near a busy road, lined with fast food restaurants, car repair shops, and dilapidated motels, as well as several boarded-up and abandoned buildings. Similar to many other parts of Stockton, the area is economically depressed and a general air of hardship pervades. While the rest of the state enjoyed the height of an economic boom, the problems of the current recession were already in full swing in this part of California. Many of the motel residents on the strip near the shop are former homeowners who lost their homes due to foreclosure.[19]

As a small-business owner and member of the chamber of commerce, Joan Davidson often finds herself on the opposite side of the bargaining table when it comes to local labor issues. Yet, she is very concerned about the economic situation in Stockton and the plight of those who had lost their homes and jobs, issues that motivate her involvement in local politics. She believes a coalition between labor and other community organizations would be a positive force in shaping public policy and feels that if the unions and the community organizations had formed a stronger, more unified coalition, they would have had a bigger impact on the political decision making around water privatization. She described her belief in the power of community coalitions and argued that "the broader the coalition the better they are and the more credibility we have if we have people coming from different outlooks joining in a group." At the same time, Davidson feels that having coalition leaders who are not part of the union brings added credibility to labor-community alliances, and argues that unions should work more closely with community coalitions because they represent the interests of the wider community rather than the vested interests of their workers and their organizations. When I asked her if she thought that labor and community organizations should work more closely together, she responded,

Oh, I do think so. Absolutely. Because if you only hear from one segment and it is considered a radical segment, like labor, for example, you tend to take only so much. You take everything kind of with a grain of salt. You think, "Well, this is their agenda, this is their program." But if, on the other hand, you have a group that is well respected for their neutrality on things—not neutrality, but independence—then there is a balance that is struck. So I think that is a good thing.

Despite a general support for labor-community alliances, Joan Davidson feels that organizations outside of the labor movement, including conservation and voter advocacy organizations, are more likely to have a broad social appeal because they are not tied to vested interests. She also feels that coalitions should reflect the middle ground, rather than what she perceives as the interests of "radical groups" such as labor. Unlike the individuals representing the community organizations, whom Davidson believes were fair and balanced in their approach to water privatization, she felt that the unions in Stockton were acting largely in their own self-interest to protect their jobs. Many other nonunion coalition members in Stockton described similar sentiments about the motivations and participation of unions in the anti-privatization fight. But community organizations and leaders played a role in shaping the development and outcome of the coalition and these perceptions expressed by nonlabor members of the coalition are revealing. They suggest that the failure of the movement to coalesce around a shared identity or argument has as much to do with the culture of the union as it does with the class-based antiunion sentiment of the community members.

Joan Davidson worked closely with the community coalition from the beginning in her role as elected official, and then subsequently as a member of the steering committee. While she had no previous ties to the labor movement, she knew many of the community leaders and considers them "well-respected citizens in this community, who were always paying attention to local government." She described the union leaders as being motivated by self-interest and regarded the nonunion members of the community coalition as "the real leaders" in the anti–water privatization movement. She saw them as "the core group of dedicated people who care, who are generally educated, generally have a stake in community, have been here long enough to know the history, and also have enough knowledge of the rest of the world." The general perception among nonunion coalition members was that the labor movement is motivated for the most part by labor market interests and therefore cannot truly represent the wider community. With this attitude from fellow coalition members and the direct threat faced by union members, it is possibly not surprising that labor withdrew from the anti–water privatization movement in Stockton.

The antiunion sentiment expressed by Joan Davidson was echoed by many of her fellow coalition members. Yet most of them worked alongside the union leaders at the beginning of the anti–water privatization movement, particularly as the unions provided most of the financial resources and organizing support needed to mount the coalition's effort. As Bernie Jacobs, a retired professional and member of the coalition steering committee, told me, "At first, all of our meetings were held at the union hall downtown. . . . They gave us free use of the room, their personnel was there . . . they contributed money, they came to the meetings, they sent representation, they got back to their members and all of that. They gave us every convenience." By providing much-needed financial and material support, as well as advice on organizing petitions and collecting signatures, the union was instrumental in the early formation and development of the coalition.[20] As time progressed, however, so did the coalition members' perception that the union was not working in the interest of the public, but for the protection of their workers. The coalition steering committee's negative assessment of the union's involvement intensified when the workers failed to show up as volunteers for the critical gathering of signatures for the referendum to overturn the city's approval of the privatization. The coalition had decided not to hire professional signature gatherers. As Barbara Smith, a key member of the steering committee, told me, "Two or three of our members of our steering committee . . . felt that was fine, because the plant employees would turn out. They didn't! They didn't! They didn't! At that point, they were dealing with their union chiefs in terms of where they were going to get the best deal. They were also disappointed and discouraged that the city had gone ahead with this project."

Barbara Smith is a committed community activist, who traces her organizing roots back to her student days at Berkeley. She is a strong believer in social justice and feels that broad coalitions are important because they "add more power, more voices" to social movements. Although she believes that movements can be more effective when unions and other community organizations work together, she feels that the union representing the plant workers was very centered on the workers' own self-interest, arguing that, "certainly unions aren't going to go against something that is going to lower their paycheck or put their future job in jeopardy." Smith explained that this view was widespread among nonunion coalition members and supporters. As a result, the coalition was strategic in presenting a clear message that they "weren't working in the interest of unions" but "in the public interest." Part of the strategy to distance themselves from the union stemmed also from fear that the mayor and council would dismiss the coalition as another "interest group." In addition, "Something interesting our attorney said is that we should be clear in our court briefs that we were working in the public interest. We weren't working in the interest of unions, we weren't working for the local water run-off organizations but we were working in the general interest of the public, which we were and we listed reasons for that," Smith explained.

Just as the union strategically chose to focus on workplace issues, the coalition intentionally framed their opposition to water privatization as representing the broader public interest and explicitly not about the interests of the workers at the plant, demonstrating that the movement's failure to coalesce is explained by both the union's internal culture and the ideological antiunion stance of the community members.[21]

The division between the union and the community members not only weakened the coalition, but also caused internal division within the coalition itself, particularly among members of the steering committee. The antiunion sentiment caused by the withdrawal of the union after privatization led many members of the steering committee to argue against working with the labor movement in the future. Others disagreed. The difference of opinion created tension within the leadership circle. Here's Bernie Jacobs on the issue:

[The unions] did come in and they contributed and they participated . . . in a very obvious way. But I think that there was, among some members, there was a feeling that although we needed the unions, we needed their money; that unions were not such a good thing, and we shouldn't give them too much support. And my position was "What are you talking about? These are human beings, these are workers, these are people who contribute to this community; by all means we should work with them—and look at all the help they give us!"

The conflict among the leadership resulted in an atmosphere that, according to Jacobs, was "not pleasant" and created lasting tensions between individuals who had worked closely together throughout the struggle to prevent privatization. He and others who supported his view were frustrated and disappointed that decisions about working with labor were made without any discussion or official motion.

The nonunion coalition members who supported the labor movement also felt that the coalition could benefit from the union's role as a key political power broker in the community, and were disappointed when their views were dismissed. As Bernie Jacobs recalled:

It wasn't ever anything that was fully discussed. A lot of things are discussed, but I just saw that some people, as I say, didn't like being between labor/management sort of thing, which I had completely disagreed with. Others maybe didn't like going to a union hall and so on. And I tried to make the case that some of these unions have several thousand members and they have their own newspapers and political clout. We can get our story to them, they put it in their paper, and several thousand people will know and be involved in this and perhaps can become volunteers. But it was, "No, no, we are not interested."

Jacobs was one of the first members of the coalition against privatization, becoming involved through a friend and colleague from his workplace before he retired. Although he attended meetings from the beginning, he had no previous ties to any of the coalition members, other than his work colleague. He joined the movement out of a strong sense of social justice. He considers himself a social advocate and felt his background in community organizing would be useful to the coalition's work. He sees a strong connection between the concerns of labor and those of other community organizations, including environmental and social justice groups, but understands the "conflict between the needs and the goals of a few people who can make a lot of profit and the needs of the more ordinary people who need to just make a living." Although

Jacobs continues to play an active role with the coalition, he is critical of many tactical choices the organization has adopted, particularly the decision to distance itself from the concerns of the labor movement. Unlike many other coalition members interviewed, Bernie Jacobs sees the movement in terms of failure as well as success, and believes that had the coalition worked more closely with unions and other social justice groups, they would have been able to prevent privatization without the legal challenge, saving the city millions of dollars.

The reluctance on the part of the steering committee to include the wider community, such as labor and social justice organizations, in the coalition, weakened its capacity to influence policy decisions, particularly around water privatization. At the same time, the reliance on traditional business unionism on the part of the plant employees and union members, and their subsequent retreat from the movement, alienated the other coalition members, and damaged the alliance between the labor movement and the community organizations.

Despite respect and sympathy for the union and its members, some coalition members continued to blame the unions and their focus on workplace concerns for giving the mayor an opportunity to neutralize the argument and divide the movement. Wendy Lawson, a member of the steering committee and a community educator, sympathized with the plant workers and their fear of job loss, arguing that it was a "scary time." Lawson, a wealthy, highly educated mother of five children, had worked closely with the plant employees through her environmental education work focusing on water quality and conservation. She told me that she knew many of the workers at the plant and appreciated the work that they did. Nevertheless, she felt that the union's message was detrimental to the success of the movement, explaining that "if your only argument is the loss of municipal jobs, and the contract says everyone who has a job gets a job, and that job is going to be equal or better than what you already have, then the union's argument [is] lost." Despite the fact that union members felt that they were working for the good of the entire community, by withdrawing from the movement and concentrating on labor market concerns, their message of community solidarity failed to resonate with nonunion coalition members. They created a division within the movement that was seized upon by political representatives in favor of privatization.

As a result of internal movement division and a disjointed argument, the coalition was unable to persuasively convince political elites to support their cause, enabling the mayor and his supporters to dismiss the opposition movement, as a "small rag-tag group of ideologues and greedy workers," one key political figure told me. Because of the rift between labor and other community organizations, a broad-based coalition required for a successful movement outcome never solidified in Stockton. The divide between the community coalition and the unions resulted in a fractured framing strategy, with the workers focusing on work-related issues and the coalition targeting the mayor's and council's views as unrepresentative of public opinion. This sharply contrasts with the Vancouver case, in which labor successfully played a central role in the coalition against water privatization.

Vancouver

In Vancouver, there were key differences in the nature and outcome of the labor-community coalition that emerged in response to the proposal to privatize the water treatment plant. The anti-privatization coalition included members of environmental groups, social justice organizations, and labor unions, including one of the largest public sector unions in British Columbia, the BCPSEU. When it became clear in early 2001 that Metro Vancouver was moving forward with plans to privatize the Seymour water filtration plant, researchers at the BCPSEU quickly began mobilizing to get the anti-privatization message out to the public.

At first there appeared to be little interest in the issue outside of the union itself, but when the research and communications team—made up of three longtime community organizers—reached out to their networks outside of the labor movement, the issue began to gain traction in the wider community. Eric Robinson, a fifty-five-year-old political and community activist with ties to other social movement organizations and political allies, was one of the three veteran community organizers who worked full-time on the anti–water privatization campaign. He explained to me that a key tactic to broaden the issue beyond the concerns of labor was to appeal to the media, to environmental organizations, and to the general public by framing the issue as critical in terms of public control and environmental protection. "The first thing we did is we got funding from [our] national [office] to do a poll. And what that found was that first of all nobody in Vancouver knew that this was being done, and when they found out they weren't happy about it," Robinson explained, describing his efforts to mobilize the public. He recalled the importance of media attention to the issue: "We were finally able to get some breakthrough on it because . . . all of a sudden after that story broke in the newspaper, people began to get a lot more interested because of the concerns they had about public control and accountability." Along with the media attention, Robinson described the importance of support from the environmental community for the Vancouver anti–water privatization movement:

So then I gave a speech [to a local environmental group] and discovered that they were extremely interested. It was a much stronger response than I had expected, and these groups were really clear that they were unhappy about this and wanted to do something about it. . . . We realized that there was a lot of interest from environmental groups. So we went down to the public library and got their list of environmental groups in the region and started doing mailings to them . . . basically if there was an environmental group that we could find in the Lower Mainland, we were sending information to them.

By mailing information to environmental groups, the union was attempting to engage them in the issue, laying the groundwork for a future coalition opposed to water privatization.

According to Eric Robinson, the support of the environmental community was fundamental to the movement's success. He described how when environmental organizations joined the cause, the issue of water privatization "stopped being a union issue and became a community issue." By involving the wider community and strategically framing the issue as a public concern—one that affects both local control of resources and environmental protection—the Vancouver-based local of the union was able to move the issue beyond the workplace and into the public arena.

One important difference between the movement in Stockton and Vancouver in terms of the union involvement is that, while the labor movement in Vancouver was heavily involved in the anti-privatization movement—in both mobilizing the public and providing critical resources—the union that played a central role did not actually represent the workers at the water treatment plant. Hence, unlike UW5 in Stockton, the BCPSEU did not focus on workplace or labor market concerns, but instead framed the problem of water privatization in such a way as to mobilize other movement sectors and appeal to the public.

The BCPSEU is one of the largest public sector unions in British Columbia, representing over 100,000 employees. A strong member base provides a strategic advantage to the BCPSEU by allowing them to shift attention away from the traditional business unionism model and its focus on organizing new members and concentrate instead on issues with broad public appeal. As part of this strategy, the BCPSEU often downplays their role in campaigns, to avoid being labeled as self-interested or alienating the broader public. Jim Roberts, the current president of BCPSEU, explained:

We try to stay in the background to avoid hurting a campaign. Sometimes as soon as people hear the word "union" it sets off this alarm in their head that, "Okay, it used to be a really good thing but now everything is wrong because the union is involved and there has to be sort of an alternate agenda somewhere." And so we can't spend a lot of time out front because a campaign would actually be over in any number of areas if we did that, even though we are always well entrenched.

The BCPSEU does not need to focus on organizing workers, which has allowed it to transition from traditional business unionism to social movement unionism, in which organizing shifts from the workplace to the community. While campaigns remain important to job protection, wages, and workplace conditions, they are also more broadly focused on the wider social good and the public policy domain. Building community solidarity is an essential strategy for social movement unionism and the mandate of the BCPSEU is reflective of this critical awareness of the importance of community support.[22] For example, the campaigns highlighted on their Web site and in their organizational documents focus only indirectly on job- and worker-related concerns, and instead concentrate on issues such as sustainable communities, local economic development, and alternatives to neoliberal global capitalism.

Building alliances with the wider community was a central tactic adopted from the outset by the BCPSEU anti–water privatization organizers. For this reason, they downplayed the "jobs" issue, including a strategic decision for the union representing the workers to stay out of the fight. "One of the reasons we got credibility on this is that we fought to keep the water public, but it wasn't our members. And I think that gave [us] some credibility because they couldn't argue we were just doing it for the workers [because] we've never represented those workers," recalled labor researcher Heather Harrison. This strategy allowed the BCPSEU to shift the focus from labor market concerns and redefine the struggle in political and social terms.

Beyond generating credibility with the public and with elites, the involvement of a union that did not represent the workers also allowed labor organizers to reach out and build a coalition of community groups from diverse social movement sectors because they were not perceived as purely self-interested. The BCPSEU had a sophisticated understanding of what their role in the movement should be: to provide vital resources and organizing power, while taking a backseat in terms of the public face of the campaign.[23]

To avoid criticism of union involvement and to strengthen the movement opposing water privatization, the BCPSEU provided the resources for other organizations and activists to lead the fight against privatization.

Jim Roberts, the president of BCPSEU, has worked his entire career with organized labor, but believes that unions cannot generate the kind of political clout that is necessary to protect public sector "living wage" jobs without building solidarity with community organizations. He feels that a key area for coalition building is with the environmental movement and told me that "the link has never been stronger." An essential component of building community solidarity is to provide resources—from research and legal opinions to public workshops and conferences—to organizations outside of the labor movement to help make linkages between the concerns of labor and those of other movement sectors. The BCPSEU researchers and organizers interviewed explained that the union regularly devotes considerable resources and educational outreach to engage communities and groups outside of the labor movement in the issues that they believe are important for the good of society as a whole. As a result of the community outreach work, the union has developed and fostered strong social ties across various movement sectors that they are able to draw upon for support as opportunities for collective action arise.

The union's involvement in the anti–water privatization movement is an example of the BCPSEU's work building truly broad-based coalitions. Eric Robinson noted that the union's success in engaging movements outside of labor is tied to their ability to argue their case in terms that resonate beyond union members. "The biggest thing was simply public control. If you had to summarize what everybody was concerned about it would have been summarized under the frame of public control of water. That was our focus," he explained. A strategic choice not to be the public face of the movement and to frame the issue as a public policy concern rather than a labor issue was essential in mobilizing a wide pool of activists and organizations from other movement sectors. When asked why he thought the movement was successful, Robinson replied:

Because it simply became such a broadly based opposition. If it had really been sort of a narrow union thing, if it had just been [the BCPSEU], it never would have been successful. . . . We were able to make it about something that matters to the public and not just workers. In this case, we were able to say, "We are opposed to public/private partnerships and are particularly opposed when it comes to privatization of water services, but we don't represent these workers. This is a policy perspective."

For Eric Robinson and other members of the BCPSEU, working with a coalition of community partners shifted the privatization from being a union issue to one that was of community concern, and allowed the campaign to take off and mobilize a wide range of activists from diverse backgrounds.

Community Coalitions: Collaboration, Solidarity, and Dense Networks
An important result of social movement unionism is the support it enables unions to generate from the broader public. The perceptions in the community of the BCPSEU reflect the union's strategy to focus on issues that would appeal to constituents outside of the labor movement. Many of the respondents from diverse community organizations spoke favorably of the union's involvement and particularly of the focus on issues that resonated widely. Many of them were individuals who had no previous ties with the labor movement, but recognized and appreciated the BCPSEU's efforts to reach out to nonunion members. Bob Davies, a seventy-year old retired engineer and member of a local chapter of the Citizens Action League (CAL), explained that the BCPSEU made important connections between the concerns of their members and those of the general public:

The opposition to the water privatization was based on the fact that it should be a public utility. I think a lot of [the] labor movement, particularly such as we still have it in Canada, is people who work in the public activity areas, health care, water systems. And their issues obviously spill over into much larger issues that concern everyone. I think you find the labor movement both in health care and this water issue has quite a significant input because they make those connections and go beyond the focus of their members' benefits.

Davies feels that, in general, social movements have demonstrated little success in creating social change. Yet he credits the success of the anti–water privatization movement in Vancouver to the ability of organizations from different movement sectors to get past their differences and work together. He said that the strong labor-community coalition in this case led him to "question [his] belief that the Left can't get it together and do something—in a unified sense coherently to the external world."

A sense of appreciation for the support of the union was widespread among members of community organizations in part because of the union's decision to provide resources and organizing capacity, but remain on the sidelines in terms of public presence. The involvement of professional social movement staff also likely strengthened the support for the BCPSEU from the wider community, as traditional class-based differences would have been less prevalent than those that tend to exist between workers and activists from more middle-class community organizations.[24] Fiona Rogers was one of the main community organizers of the anti–water privatization movement. She is a thirty-seven-year-old community educator, who has been active in issues of globalization and social justice since being an undergraduate in Montreal. After graduating with a BA in political science, she completed a master's degree in environmental studies, with her research focusing on environmental justice movements in Latin America. She also spent a year in South America, working with indigenous groups fighting to prevent privatization of their water systems. At the time of the proposed privatization plan, she was working for a local theater company that, coincidentally, was mounting a dance theater production about water privatization. She was hired to "build community involvement around the project and . . . to connect with other people doing water work." When the issue of water privatization came up, she began to focus some of her outreach work on "activism around the issue" and ended up working closely with both the BCPSEU and the BC regional office of CAL.

Fiona Rogers explained that the ability of the BCPSEU to provide resources without being a visible movement leader was critical to the movement's success:

The [BCPSEU] was really good at not stealing the limelight, which is really important because unions often do that, and it pisses off community people to no end. But they were really supportive, and also strategic around realizing that if they were too visible then antiunion people wouldn't pay attention to us. And so there were a number of times where they had either financed something or really played an important role research-wise, and they would say, "Don't put our name on it. Don't make this a [BCPSEU] thing."

A strong understanding of the importance of connecting with community, and balancing resource and organizing support with the need to keep a low profile was instrumental in building trust and credibility with organizations and individuals outside of the labor movement. As a result, a unified opposition movement was able to present a strong and convincing argument against water privatization.

The anti–water privatization movement in Vancouver was not the first time that the BCPSEU had worked with organizations outside of the labor movement. Although their primary focus is on protecting public sector jobs and services, they have a long history of working with community groups to raise awareness of the importance of public services and utilities. Several years before the water privatization issue came up in Vancouver, the BCPSEU had forged alliances with other organizations across the country—including CAL—to raise awareness of the implications of water privatization for communities. The focus on community issues enabled the BCPSEU to connect labor concerns with those of the wider community. "Water was a huge environmental issue for our organization and how the international trade agreements were impacting everything," recalled BCPSEU president Jim Roberts, describing how the issue of water was critical for the union to make connections and building networks with other organizations. He told me that water gave the BCPSEU "the perfect kind of footing for us to get involved and work with other people and talk about why we need to keep water public." When he began researching water privatization, he quickly realized how it "tied to a whole lot of other things—industry, small business—besides the sort of big platform issues that we were talking about." Roberts was shocked at the "the implications of what losing control of water might do to not only the community but an economy" and became aware of the importance of building coalitions with other organizations and movements focused on protecting water.

The potential broad-based appeal and multidimensional issues around water privatization facilitated the union's links with other important organizations and interests. Despite the fact that water privatization would not directly affect the workers they represent, the BCPSEU believed it was in the long-term interest of their workers to participate in the campaign against water privatization.

The networks that formed out of previous work around water were also strengthened by the battle over the Multilateral Agreement on Investment (MAI), which a coalition of labor unions, environmentalists, and social justice groups had worked to defeat the previous year.[25] Many activists in Vancouver described the importance of the anti-MAI coalition for the ability of the water movement to organize quickly and effectively against water privatization. The relationships that had formed among diverse organizations and the educational outreach around complex issues of trade and investment provided the foundation for the anti–water privatization movement to emerge and develop as a cohesive and broad-based coalition, with a shared collective identity. Many of the Vancouver activists, including Mike O'Brian, referred to the MAI movement as being an important precursor to the anti–water privatization coalition because it united movements around the dangers of trade agreements for local control of resources; having raised awareness about trade concerns with the public made it easier for organizations to draw parallels with the issue of water privatization and mobilize a wide pool of constituents. O'Brian, a water campaigner for CAL, explained the importance of the preexisting network that formed during the stop the MAI campaign:

That is where the network from the MAI came in really handy. They had the grassroots organizers and the networks established, even down to where to have the meeting. They knew each other, they had worked together, they had the success of the MAI and that was seen as a huge victory. It was very broad-based; it was environmental issues; it was anti-privatization issues, poverty issues; it was labor groups, globalization groups. Water then became kind of the next battle, as opposed to reinventing the whole process.

The MAI networks in Vancouver were essential for creating a successful movement not only because of the strong ties among labor unions, social justice groups, and environmentalists, but also because of the preexisting ties between the activist sector and political elites. While social embeddedness among community organizations matters for movement outcomes, so does the presence of sympathetic political allies who provide opportunities for grievances to be heard.[26] In Vancouver, preexisting ties among activists and political representatives facilitated access to key political opportunity structures that were critical to shaping movement outcomes.

Unlike in Stockton, the movement in Vancouver was able to create a successful labor-community alliance because the union involved in the coalition constructed their opposition in terms that would resonate with movements outside of labor. This strategy—based on social movement unionism—was important because it allowed the union to reach out to their networks and provide critical resources while remaining in the background so as not to make it appear a union issue. Beyond the culture of the union and the preexisting networks, the coalition in Vancouver was successful because of the density and cohesion of the organizational field as well as the vital role of a frame-bridging organization.

Organizational Density and Social Cohesion: Bringing Together Networks and Frames

As I have just described, the role of coalitions for mobilization and movement success and the dense networks among diverse movement organizations in Vancouver were crucial in facilitating the emergence of a strong and unified anti–water privatization coalition. The existence of strong ties among movement sectors can shed light on either the presence or absence of coalitions or their relative strength, but how do these coalitions form? Understanding this is useful for explaining why similar movements take different trajectories in terms of network structures. What contributes to the creation of cross-movement coalitions?

One of the main differences between the movements in Stockton and Vancouver is the density and cohesiveness of the social movement organizations in Vancouver. Beyond the culture of the union and preexisting networks, the existence of dense ties between labor, social justice, and environmental organizations suggests a high level of social cohesion between diverse movement sectors before the coalition against water privatization was formed. Of particular importance in the network of organizations was the centrality and connectedness of a prominent social justice organization—the Citizens Action League (CAL), not surprisingly, mentioned throughout this book—with ties to both the labor movement and the environmental movement. CAL played a central role in bridging the green-labor divide and creating strategic alliances across movement sectors. CAL's main focus is on social, environmental, and economic issues, such as fighting to protect public services and ensure environmental protection. Because of this broad concentration, CAL often finds itself working closely with both labor and environmental groups. In 2001, when the issue of water privatization in Vancouver came to the forefront, CAL's BC office had been working regularly with both the BCPSEU and environmental organizations on a range of issues facing the region, including the concern over bulk water exports and the efforts to stop the MAI. Many of the activists interviewed in Vancouver were connected to CAL, either through existing alliances, membership, or volunteer work. Figure 6.1 reveals the density of overlapping ties between activists from environmental, labor, and community organizations. The graph reveals the centrality of CAL in the dense cluster of social movement organizations and activists and shows the overwhelming connections CAL has to the diverse activists. The size of the triangle represents the centrality of the organization, with CAL being the largest and hence most central organization in the network of anti–water privatization activists and organizations.

Many activists spoke of the importance of CAL for bringing organizations together and connecting movement agendas. Lynn McCain, a fifty-six-year-old office administrator and vice president of a BCPSEU local spoke about the important relationship that existed between the union and CAL, especially with the work they had done around water issues and concerns. She described their relationship as strong, explaining their history of working closely together: "We always have had that relationship because they are really big into the water stuff. We've done events with them, and we support a lot of their ideas and we tell each other—information sharing. We have that connection." In order to mobilize people to join the anti–water privatization movement, CAL and the BCPSEU teamed up to organize several public information sessions, with the goal of reaching out beyond their own membership base. As union researcher and community organizer Heather Harrison explained, CAL was instrumental in connecting the BCPSEU with a wide range of individuals and

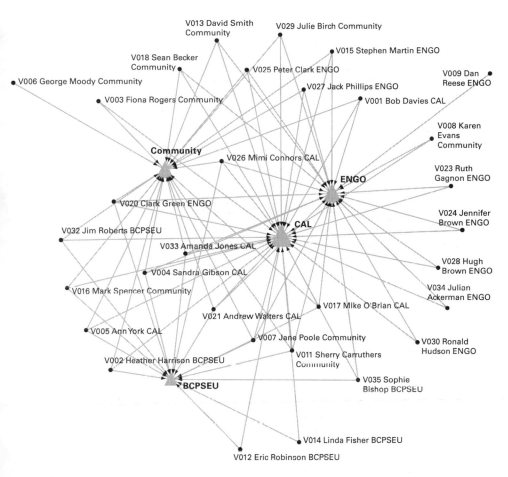

Figure 6.1

Network density in Vancouver. The graph shows the density of overlapping ties between Vancouver activists from environmental, labor, and community organizations and reveals the centrality of the Citizens Action League (CAL) in the organizational field. The size of the triangle represents the centrality of the organization, with CAL being the largest and hence most central organization. CAL—Social Justice Organization; ENGO—Environmental Organization; BCPSEU—Labor Organization; Community—Community Organization groups across the Lower Mainland. She described the importance of CAL as a bridging organization:

When you get to working in coalitions, the [Citizens Action League] was key . . . They helped us connect not just with their members, but with many other groups. I remember that at one of the public meetings a woman stood up and said something like, "I'm your worst nightmare. I'm educated and I'm recently retired and I have nothing but time on my hands." And we had no idea who she was or who most of the people were. So, we knew that this thing was going to be big. We thought we were going to win when we completely lost control of the campaign! So it just took off, it just resonated with people. The [Citizens Action League] was key with that.

One of the reasons that CAL was so effective at mobilizing a large number of people to join the movement was linked to their membership base, their recent history of winning campaigns and their internal structure that gave them access to critical resources. "CAL was a force to be reckoned with. We had some major national victories. We had contributed to stopping the MAI, we contributed to stopping the merger of the banks, we had stopped the cutbacks on the old age security. So we had some major clout," explained Andrew Walters, an activist and past chair of a local CAL chapter. He described the fundamental role played by CAL in ensuring the movement's success both in terms of their credibility in the community and their access to a large membership base: "At that time we had just opened the provincial office and the local organizer was very effective. We worked together on mobilizing [CAL] members of whom there are probably well over 5,000 in the region."

CAL's access to a large and diverse membership, made up of individuals across the Lower Mainland, was critical in ensuring a significant turnout at the public consultation meetings organized by the Metro Vancouver Board. The ability to mobilize an extraordinarily large turnout at these public consultation events turned out to be one of the most important organizing outcomes for the movement because it was a signal to the political leadership that there was widespread public opposition to their proposed privatization plan.

The structure of CAL also facilitated their ability to mobilize resources and reach out to diverse networks, acting to connect a variety of groups around a common cause, as Mike O'Brian, CAL's water campaigner, described:

At the time, we had local organizers in a few cities across Canada, Vancouver being one of them. That was really effective in taking part in the anti-privatization because there were a lot of people already following it through an activist network, and we had a full-time organizer to assist with that. A lot of people who were involved, were involved in other things that were connected. You know, same old gang, same people, different meeting. It had a core of people who make the connections easily—especially those who work with the [Citizens Action League]—and then reach out to their networks, like labor groups.

CAL's dense organizational network structure, including its many connections with local political opportunity structures facilitated the sharing of information and mobilization across diverse organizations and movements in Vancouver. The anti–water privatization activists in Vancouver were from multiple social worlds, but through existing network ties and previous campaign work, moved easily between worlds and worked collaboratively together.

Organizations such as CAL are important not only for network integration, but also for synthesizing the goals and strategies of different organizations to construct common frames.[27] In Vancouver, activists were able to draw on a common master frame—the risks to public services from corporate power and international trade agreements—that resonated across movement sectors and organizations. The central role played by CAL facilitated the successful framing strategy because of the organization's campaign emphasis on corporate power structures and international trade agreements, which easily linked the concerns of both environmental organizations and labor unions through the identification of corporate hegemony as a common root cause for deregulation of both environmental protection and public services. Mike O'Brian described the fears around the corporate agenda as being the common connection across movement sectors: "That particular movement brought a lot of different analytical perspectives into the fight—right from poverty issues to trade— making that corporate agenda connection very strongly. . . . So all of these organizations that were fighting different issues, but with the same analysis, found it very easy to come together over water. What [CAL] did is inform people about the corporate agenda and made those connections." The focus of CAL's campaigns shed light on the connections between corporate and trade issues and the concerns of diverse organizations, allowing for the development of a common frame around the corporate control of water that resonated across both the labor and environmental movements.

The BCPSEU had completed in-depth research on the detrimental consequences of water privatization and the effects on water quality, public accountability, and transparency, with a particular focus on the risks under NAFTA's investment clause, including commissioning a legal opinion from one of Canada's experts in international trade and investment law. The focus on corporate control and trade was instrumental in connecting their agenda with that of the Citizens Action League. Where the union had less success was in merging its analyses with the concerns of environmentalists, whom they needed to mobilize because of their significant clout in the community and their access to decision-making arenas. Connecting the labor and environmental movement was an essential strategy in building a broad-based movement in Vancouver that would resonate not only with members of the general public, but also with political elites. CAL was instrumental in bridging the green-labor divide by linking together environmental and labor concerns under the common master frame of corporate power.

Jennifer Brown, an environmental activist and chair of the locally based group Conservation Now, was one of the leaders of the coalition against water privatization. Because of her expertise in issues of water quality and her role as chair of a prominent conservation organization, she was appointed as a citizen representative on the Metro Vancouver Water Committee, a position she filled for several years. It was important that the coalition have the support of environmental leaders like Brown, not only because of her expertise around water issues, but also because of her strong connections and credibility with local policy makers. She described the role of the Citizens Action League in bringing together environmentalists with labor representatives:

[Conservation Now] played a somewhat larger role than you would expect a local conservation group to play, and that was in part because of our good links with our local chapter of [Citizens Action League]. . . . The other was this rising awareness of how important public assets were coming under the control of private interest for profit. And I really credit the [Citizens Action League] with doing a great job with bringing in the corporate issue. We were lucky because one of the key members on my conservation committee was also very active with the local chapter of the [Citizens Action League] and so it ended up being a really good synergy for us.

During the interview, Brown explained that CAL connected Conservation Now with the BCPSEU by organizing a meeting with environmentalists and labor activists around the theme of trade and corporate control. The focus on trade issues and the connection to environmental protection was instrumental in securing the support of key environmental leaders in Vancouver.[28]

Amanda Jones, a forty-year-old mother of two who worked as CAL's regional organizer at the time of the anti-privatization battle argued that

CAL not only worked to bring together different organizations, but also focused on creating a common messaging strategy to synthesize the concerns and goals of environmentalists and labor unions. "Because [CAL] really worked from the trade perspective and privatization issues were major themes . . . our materials focused on that," she explained. Jones went on to say, "We were able to show how . . . what the [BCPSEU] brought to the table, what the environmental groups were doing . . . all reinforced each other and when we had meetings we would talk about making sure that we were reinforcing each other's messages and not contradicting each other." CAL's role was no accident. They strategically acted to build on the strengths of each movement's potential contribution and sought to bring them into the coalition.

In successful social movements, coalitions facilitate the shift from single-issue grievances to broad common agendas, widening the pool of movement supporters, creating a sense of solidarity, and generating a unified movement that cannot easily be dismissed by elites as representing the narrow interests of a few organizations.[29] In Vancouver, organizations from different movement sectors worked together along several strategic dimensions to build a cross-movement coalition to fight the privatization of water. Coalition partners shared resources, used agreed-upon organizing tactics, cooperated across networks, and adopted common frames that brought together the concerns of diverse movement sectors.

Stockton: Network Fragmentation

The level of network density and social cohesion is a key explanatory difference between the movements in Stockton and Vancouver. In Stockton, despite the existence of multiple social movement organizations, these groups were fragmented, with few existing ties between organizations, particularly environmental and community groups and labor unions.[30] In addition to network fragmentation, the network profile of coalition members in Stockton also mattered. In Stockton, coalition members were either associated with labor unions or community organizations, with very little overlapping membership. Figure 6.2 presents respondents membership ties and reveals the low network density and integration across movement organizations among Stockton respondents. The graph demonstrates that the organizational field of environmental, labor, and community organizations was much less dense than in Vancouver, with little cohesion between social movement organizations and actors involved in the movement. In particular, there were few ties between UW5 (the main labor organization) and other organizations, including environmental

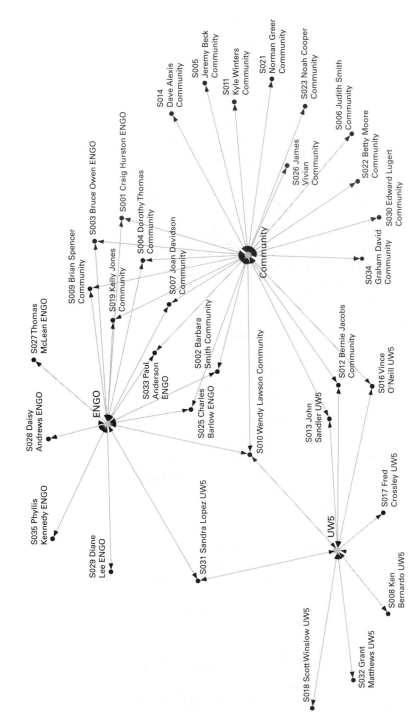

Figure 6.2
Network density in Stockton. The graph shows the lack of density and integration between social movement organizations in Stockton. The graph demonstrates that there were few ties between the Utility Workers Local 5 (UW5, the main labor organization) and other organizations, including environmental and community organizations. Coalition members were either associated with labor unions, environmental groups, or community organizations, with little overlapping membership. ENGO—Environmental Organization; UW5—Labor Organization; Community—Community Organization

and community organizations. Coalition members were either associated with labor unions, environmental groups, or community organizations, with little density and integration across movement organizations.

The movement in Stockton also lacked the facilitating role of a central organization—like CAL in Vancouver—and thus it remained a loose network of organizations and individuals unable to coalesce around a common frame. There were few preexisting networks between the organizations that formed the anti–water privatization coalition. The main organizations besides the labor union—including a prominent conservation group and a high-profile voter advocacy group—were traditionally unsupportive of labor concerns. Despite the initial alliance between unions and other community organizations at the outset of the Stockton movement, the lack of social cohesion between the groups meant that the coalition remained divided around both tactics and frames, with unions advocating a focus on workers and job protection, while other coalition members focused on accountability and democracy. As a result, it was easier for the movement to be dismissed by political elites as representing only a small group of self-interested individuals.

Conclusion

The contrasting examples of the Stockton and Vancouver movements demonstrate that an important factor in explaining the success of anti–water privatization movements is the development of broad-based, unified coalitions that create consensus over tactics and frames. The activists interviewed in Stockton and Vancouver spoke of their understandings about coalition building in the face of a common grievance. Although their answers reflect a wide range of opinions and sentiments, activists from both movements described the importance of coalition building and community solidarity for successful movement outcomes. Yet the movement in Vancouver demonstrated a more cohesive and interdependent alliance between labor unions and community organizations than the movement in Stockton. Why?

First, dissimilarities in the culture and structure of the labor movement led to differences in both meaning construction and the role played by unions. In Stockton, the union that was centrally involved in the coalition opposing water privatization—Utility Workers Local 5—had an organizing structure shaped by a business unionism model, the focus of which rarely extends beyond labor market concerns. As a result, the union members involved in the movement framed their grievances around job protection and wages, a strategy that was resisted by other members of the coalition. By concentrating on issues of self-interest, the union was unable to build solidarity in the wider community, ultimately leading to a division between the union members and the other coalition partners.

In Vancouver, the role of the labor movement took on a different form. The BCPSEU, the main labor union involved in the anti–water privatization coalition, was more activist-oriented and explicitly focused on building solidarity with other movements, to the point of being willing to provide resources without recognition and strategically obscure their own role for the good of the larger cause.[31] The BCPSEU has a history of framing issues more broadly in order to generate support from outside the labor movement, and in building critical alliances with organizations across movement sectors. Using frames that resonate with a wide range of organizations allows social movements to move beyond local, single-issue politics to a more thorough critique of economic globalization.[32]

The focus on social movement unionism facilitated the creation of a unified coalition that developed a shared collective identity that coalesced around the common master frame of corporate power. The labor movement in Vancouver constructed its opposition to water privatization in political terms—as a social justice issue rather than as a workplace concern—allowing it to forge a stronger, more viable community coalition than the labor movement in Stockton. The common vision articulated by the Vancouver activists reflects a shared understanding of the root of the problem and the sense of trust and reciprocity among the different groups involved in the movement. This is largely the result of the more activist structure of the labor movement.[33]

Second, the differences in coalition outcomes in Stockton and Vancouver can be explained by the density and social cohesion of the organizational fields. In Vancouver, the central role played by the Citizens Action League facilitated the merging of frames between dense network clusters—the labor and environmental movements—and the formation of a unified coalition to counter water privatization. The centrality of CAL functioned both to construct broad-based arguments that linked labor issues to wider global social, environmental, and economic concerns, and to utilize the research and analysis provided by labor to connect diverse movements together under the common frame of corporate power.

In Stockton, on the other hand, the absence of network density and social cohesion between organizations and leaders meant that that movement was unable to coalesce around a common grievance or master frame. The absence of a unified coalition allowed the political structures to neutralize the labor argument about jobs and working conditions, leaving the union with little leverage to fight the privatization of the water treatment plant, and further fracturing the movement coalition.

Research on network processes in social movements points to the importance of social cohesion and network density as a key causal mechanism in uniting movements through common frames.[34] Research that specifically focuses on coalitions demonstrates that successful movement outcomes are dependent on the ability of diverse movement sectors to build interorganizational alliances and these alliances are more likely to occur when there is a high level of social cohesion between organizations or individuals.[35] Yet, while research on coalitions has identified social cohesion as a key causal mechanism for synthesizing frames and facilitating resource interdependence between organizations, it is important to go beyond a structural analysis of network mechanisms to understand *how* they operate as a process.

Beyond the density of the organizational field and preexisting ties between diverse organizations—it is important to consider the *nature* of organizations that have a high degree of network centrality and act as frame brokers, such as CAL in Vancouver. In the case of counter-hegemonic movements that depend on a unified alliance between labor unions and environmentalists—including anti–water privatization movements— the presence of *social justice*–oriented organizations is critical because of their ability to bridge the interests of environmental and labor groups. Social justice organizations play a key role in bringing together the environmental and labor movements because they have relationships with both movement sectors and thus are able to link movement frames by dovetailing environmental and social justice issues with global issues such as trade and corporate ideology and agendas.

Labor-environmental coalitions can be a potentially powerful force against privatization and the increasing commodification of the global commons as well as in fighting the deleterious consequences of neoliberal globalization. Yet, the success of labor-community coalitions around the privatization of public services depends on the capacity of labor unions to move beyond narrow labor market concerns and offer a more broad public understanding of privatization. For the labor movement, this means investing more resources in research and educational outreach to create linkages with social, environmental, and economic concerns that lie beyond the direct interests of union members. At the same time, the organizational culture of community groups also matters for the success of these types of coalitions; environmental and community groups must be willing to confront some of their members' antiunion biases and tap into the potential resources and support from labor unions. In the Vancouver case, for example, the environmental groups were more grassroots and less institutionalized and, as a result, more flexible to deploy new frames and tactics. In Stockton, on the other hand, the community groups—the main environmental organization, in particular—were formal and professionalized and thus less willing to shift focus and adopt new tactics.

At the same time, there is a need to refine and clarify the conceptual understanding of social movement unionism to distinguish between unions that reach out to the wider community for support on specific issues and campaigns and those that offer a more radical vision for social change and strategically work with the wider community over the long term to achieve this goal, as the BCPSEU was able to do.[36] What matters for regaining political power and building strong, broad-based coalitions is not necessarily mobilizing public support for labor issues, but rather the ability of unions to offer concrete prescriptions for structural change, instead of mere "perturbations" or "adjustments" to capitalism.[37] For labor to regain the political and organizing power they once held in the United States and Canada, it needs to adopt a more focused and strategic form of social movement unionism, including assuming a broader political and social agenda. Transitioning to a more activist-oriented labor movement that focuses on social transformation and provides alternatives to capitalism will create a stronger, more unified political force for labor to counteract the hegemonic power of neoliberalism.

To reverse the devastating social and environmental consequences of capitalist economies, the environmental movement and the labor movement must recognize the common root causes of both labor injustice and environmental degradation. The globalization of neoliberal capitalism provides the political opportunity structure for labor-environmental coalitions to build the counter hegemonic movement necessary for creating viable alternatives to the current economic system. Labor unions have the potential to play a critical role in movements that are less directly concerned with labor market issues, particularly those that resist global corporate power—water privatization being a key example. Social justice organizations, such as the Citizens Action League, are fundamental to facilitating labor-environmental alliances because of the brokerage role they can play in highlighting economic globalization and corporate power as the common cause of both labor and environmental injustices.

Labor and environmental movements must demonstrate a cultural shift internally to move beyond identity politics, broaden their focus, and create viable alternatives to economic globalization. This transition would require putting the traditional "jobs versus the environment" conflict to rest and focusing on what policies would counteract the deleterious effects of global capitalist hegemony, to overcome the traditional dichotomy between middle class antiunionism on the one hand and the narrow self-interest of labor on the other. Labor unions need to embrace broader social, economic, and political issues and build community solidarity to overcome "do-nothing unionism."[38] At the same time, the environmental movement needs to shed its narrow focus on "middle class" issues and recognize the link between poverty, inequality, and environmental justice if it wants to go beyond symbolic actions and effect fundamental change.[39] When labor and environmental organizations overcome their differences they create the potential to shape policies for protecting both workers and the environment.[40] Social and economic transformation and the development of new policies and institutions that reflect social, environmental, and economic justice will not be achieved by the work of a single social movement organization or sector alone. As Brian Obach argues, "[T]he creation of a just and sustainable economy depends on the ability of these two social movement sectors to work together to advance this common goal."[41]

7

Conclusion: Local Social Movements in an Age of Globalization

Perhaps the most important policy issue of our time is the question of who has access to water and who controls that access. As the world's water crisis worsens in the wake of the governance failure during the twentieth century, policy makers around the world must grapple with scarcity, pollution, and unequal distribution of water resources. In an era of globalization—marked by declining investment in infrastructure and services, the changing role of the state, and the emergence of powerful multinational water companies—local governments are increasingly under pressure to outsource water services to the private sector. The growing commodification of water has spurred the mobilization of anti–water privatization movements in communities around the world, in which activists challenge both the notion of water as an economic good and the ability of the private sector to deliver resources equitably. While there has been considerable scholarly attention to water governance and the management of water resources—including research on institutional transformations, policy reforms, and the debate between private and public sector service provision—the movements that have emerged around the world to protest the commodification of the water commons are an important and understudied subject for research.

To fully understand the conflict over resources—especially life-sustaining resources such as water—and the pathways for the development of equitable and just water governance policies, the mechanisms that underlie these movements must be a central area of investigation. As the intense and emotional narratives of water described in this book testify, water is more than an instrumental good; notions of power, meaning and identity, connectedness, and belonging permeate our understandings of this precious resource. Achieving equity and sound governance over water resources requires comprehending people's complex and multiple understandings of water and why they mobilize to protect water in their communities.

The stories of the anti–water privatization activists in this study reveal how multiple mechanisms and processes operate in combination to produce mobilization emergence, development, and outcomes. The findings suggest that the interaction of different factors is important for explaining the divergence in movement trajectories between the two cases, including cognitive mechanisms, such as ideology and frames; external mechanisms, such as the political, economic, and institutional context; and relational mechanisms, including the role of social networks and movement coalitions. When examined individually, none of these factors provides a complete explanation of differences between the anti–water privatization movements in Stockton and Vancouver. Instead, these mechanisms interact to produce variations across the two movements. Despite similar ideational understandings of water, divergence between the two movements is explained by differences in frames, political process, and social networks, as well as differences in how opportunities were seized upon by activists to create new openings for the consideration of movement claims.

The research findings also advance the understanding of how local social movements respond to globalization by demonstrating the complex pathways by which global forces shape contention on the ground. Globalization provides new problems, frames, and targets for local movements to appropriate. The success of these movements depends on the ability of activists to create *local solidarity* by seizing these opportunities, building broad coalitions, and drawing attention to the local risks from international institutions. Global shifts in capital, investment, and institutional power mean that even an issue as locally bound as water is altered by transformations at the global level. The struggles for water are both deeply rooted locally and embedded in global economic and political structures. Understanding how the local and global interact is critical for explaining the struggle over water.

Cognitive Mechanisms: From Ideology to Collective Action Frames

Environmental sociologists and water scholars point to the importance of place-based understandings of nature for shaping attitudes, values, and mobilization concerning the environment. The socially constructed and contextually mediated meanings of water articulated by activists in Stockton and Vancouver demonstrate their importance for shaping people's political understandings of water and their participation in the anti–water privatization movements in their communities. In Stockton, activists' narratives about water reflect the geographic, historical, and political context of drought, delta pollution, and water diversion proposals. In Vancouver, activists' meanings of water were closely connected to their attachments to nearby watersheds and their desire to protect what they perceived to be a clean and natural source of water.

These ideological processes help explain the emergence of environmental movements and the political mobilization to protect the commons. In the case of water privatization, these socio-natural relationships are particularly important because of the deep attachments that we have to water as a source of life and a human right.[1] The responses of activists in Stockton and Vancouver demonstrate that water is linked emotionally and spiritually to symbolic meanings and discourses of power and rights. Understandings of water are also intricately linked to "locality": the unique geographic, historical, and political context.

Investigating social-psychological and situated perceptions of water is important for explaining pre-mobilization attitudes and values that influence our decisions to participate in broader environmental movements and helps clarify why we mobilize to protect water resources as part of the commons. While social movement scholars have studied how activists and organizations *collectively* formulate and deploy frames—the shared understandings of the problem and potential solutions—in order to have grievances heard and succeed in their goals, the individual ideologies of social movement actors has received less attention. These processes should be examined separately to shed light on their distinct roles in the cycle of mobilization; ideology explains why individuals join a movement, while frames reflect the shared understandings of a collective unit, at later stages of a movement cycle.[2]

The narratives of activists in Stockton and Vancouver demonstrate that pre-movement meanings of water—as a geographically bound resource that has profound meaning in the day-to-day lives of individuals—are critical for shaping mobilization around water. But that is only one part of the story. While the movements in Stockton and Vancouver shared similar pre-mobilization ideological understandings of water as part of the commons, the frames constructed at later stages of mobilization revealed significant differences. In Stockton, the movement concentrated on issues of voter rights and democratic accountability. Activists downplayed the global nature of the problem and instead chose to frame the issue in terms of local political accountability and trust. On the other hand, in Vancouver, activists focused on the link between global and local problems, and deployed frames that centered on the threats to local accountability and control over resources from global institutions and capital. As a result, the movement was able to bridge the divide between activists and elites by focusing on community solidarity and well-being.

One of the reasons the movement in Vancouver was successful at formulating and deploying global risk frames was the presence of *global connectors*—activists and organizations connected both vertically to transnational organizations and horizontally to other activists at the local level. Most research on global social movements examines transnational activism and reveals how frames are transposed and transformed through a "scale shift" from the local to national or international level.[3] While previous research has identified the importance of vertical connections between global and local movements and frames, of equal importance is the *horizontal* diffusion of frames linking global and local problems. The connection between global processes and local outcomes has been made clear, yet an explanation of the concrete mechanisms that underlie these linkages is absent from existing research. My research focuses on the reverse scale shift—from the global to the local level—and demonstrates the specific pathways by which local activists synthesize global and local frames. The ability of local movements to construct global risk frames that resonate on the ground and create community solidarity is dependent on the presence of global connectors—individuals who are rooted in their communities as well as to networks of global activists—who help translate the "global" problem into a local threat and make sense of complex international issues on the ground.[4]

Political Process: A Multi-Scale Analysis of Social Movements

Activists in both places also responded to divergent political conditions; the movement in Stockton targeted local political institutions, while the movement in Vancouver responded to both local and global political opportunities. The Vancouver case demonstrates that activists who respond to multilevel political and economic opportunity structures create favorable conditions for influencing policy outcomes. In the broader context of neoliberal globalization and the increasing power of transnational economic institutions—including corporations and international trade agreements—a multilevel opportunity-structure analysis is important for understanding the interplay between global and local processes and the creation of new opportunities and targets for mobilization on the ground.

International power structures have varying influence over governments and resources, depending on the level of government and the nature of the resource, affecting the way in which power is constituted and resisted by social movement actors. Environmental resources, especially those considered as part of the commons, should be examined in a different light than resources that are considered commodifiable because differences in how social movement actors view the commons versus commodities shape movement opportunities, targets, and frames. In particular, the critical role of government as protector of the public realm is essential for influencing the way in which anti–water privatization activists seize opportunities and choose targets.

It is also necessary to recognize the distinct nature of national and local-level movement dynamics, particularly regarding the capacity of governments to make policy decisions in the face of the growing power of international economic structures as well as how social movements respond to neoliberal globalization. The responses of the anti–water privatization activists reveal the importance of incorporating economic structures and targets into a broader understanding of the role of opportunity structures for social movements within the context of increased economic globalization. At the same time, targeting *local* political structures is critical to these movements. Whether the context for mobilization is local or global factors into how economic opportunity structures are seized and utilized by social movement actors. Activists opposed to water privatization at the local level do indeed target international opportunity structures, including multinational water companies and international trade and investment agreements; yet the main target for mobilization remains *domestic* political structures because of the nature of both the resource and the opportunity structure.

While multinational corporations and international trade and investment treaties shape both policy decisions about water resources and mobilization by activists on the ground, actors in the anti-privatization movement use these global economic opportunity structures as *symbolic targets*, while local political structures remain the central target of contention. In this context, because the privatization of water exists merely as a threat and not a reality, activists focus on attempting to sway the decisions of political elites and influence their behavior, rather than on changing corporate policies and practices. What matters is not whether the authority of the state is waning, but how and to what extent globalization—including flows of capital and shifts of regulatory power to the international realm—has altered the capacity of the state to regulate services and environmental resources and transformed the ways in which domestic movements operate. In the case of public resources that are considered non-commodifiable, such as water, the power to regulate, conserve, and protect these resources as part of the commons is understood to be in the hands of the polity. This is especially true at the local level where the link between government regulation of resources and the delivery of services is more clearly understood than with other levels of government.

The authority of nation-states may be constrained by economic globalization and the power of international financial and regulatory bodies; local governments, on the other hand, have greater capacity to resist such pressures. As local governments are not signatories of international trade agreements, they are not beholden to international economic pressures in the same way in which national level governments are. The existence of more participatory forms of deliberation at the local level and the greater access to elites provide concrete targets and clear channels for social movements to influence policy and strengthen local political power in the face of economic globalization.

The interplay between economic and political opportunities differs depending on the political context and nature of the "commodity," and it is important to understand these differences to advance the political-economic process perspective developed by David Pellow, especially in the case of local movements resisting neoliberal globalization. Pellow argues that transnational movements are increasingly targeting institutions outside of the polity, including multinational corporations and international financial institutions, because these structures have growing power over nation-states and their ability to regulate and enforce environmental and social policies.[5] While this is certainly true for many global social movements, in the case of mobilization around resources that are considered part of the commons, social movement organizations and actors are more likely to target *political* structures rather than corporations because of the "public" nature of these resources. For the activists interviewed as part of this study, water is a local, geographically bound resource that is part of the commons, and therefore decisions about how to regulate, protect, and deliver drinking water to local populations is considered to be the responsibility of municipal governments.

Cross-Movement Coalitions and Local Counter-Globalization Movements

Differences in the nature and strength of coalition building in Stockton and Vancouver also explain why the movements in each place diverged. In Stockton, despite the presence of a wide range of organizations and activists, a strong coalition never developed. The coalition that initially formed eventually fractured, in part because of the organizational culture of the groups involved, but also due to the low organizational density and the absence of a frame-bridging organization. Both the business union-ism model of the UW5 and the more conservative nature of the community and environmental organizations in Stockton prevented the coalition from becoming a cohesive unit. The union remained focused on job and workplace issues, while the environmental and community groups were less willing to be flexible and reach across movement boundaries. The low density of the organizational field and the absence of a frame-bridging organization also meant that anti–water privatization activists—most of whom were not previously connected—did not unite together under a common frame.

In Vancouver, on the other hand, a strong and effective coalition rep-resenting a broad range of organizations—including previously discon-nected environmental and labor organizations—formed and developed a common frame. Organizational culture—including the embrace of social movement unionism by the BCPSEU and the grassroots model of the en-vironmental and community groups—meant that the organizations in the coalition were more flexible and thus able to overcome differences. The strong organizational density also mattered because many of the organi-zations and activists had worked closely together in the past. At the same time, environmental groups and labor unions, which had little experience working together, were united together under a common frame of com-munity solidarity in the face of global threats through the critical bridge-framing role of CAL, a social justice organization. Such organizations play a critical role in bridging green-labor concerns because they have ties to both movement sectors, and thus have the capacity to bridge networks while synthesizing diverse environmental, labor, and social justice frames with global issues such as trade policies and corporate ideology.

The stories of the anti–water privatization movements in Stockton and Vancouver reveal the potential for building and strengthening commu-nity coalitions in the context of countering the deleterious consequences of globalization. Movements focusing on economic, social, and envi-ronmental concerns—particularly in the context of economic globaliza-tion—have the potential to unite actors and organizations from diverse sectors, including environmental and labor movements. The success of these counter-hegemonic movements depends on the formation of cross-sectoral alliances and the creation of common frames.

Broad-based community coalitions, including labor-environmental co-alitions, are a potentially powerful force against the growing encroach-ment of private capital on the global commons. Yet, for these coalitions to be successful in offering viable policy alternatives to economic globalization, they must move beyond traditional identity or issue-based politics and offer a more broadly framed understanding of privatization and neoliberalism. Strong cross-sector alliances in the context of neoliberal globalization have the potential to overcome the constraints of identity or issue-based politics and create a common collective identity. There is growing recognition that in the face of economic globalization many social movement sectors and organizations are beginning to shift from identity or issue-based politics to a wider focus on social justice and social change.[6] Broader issues of social and economic justice have the potential to bring together previously disconnected movements and create a common collective identity in response to neoliberal globalization because they create an overarching frame that resonates widely across movement sectors.

One avenue for building stronger alliances between labor and environmental movements and creating a shared collective identity lies in *localized* resistance to globalization. Local manifestations of global problems—such as the attack on public sector services and the economic risks from international corporations and trade agreements—have the ability to forge solidarity between previously disconnected or conflicting communities such as labor and environmental movements because they enable the focus to shift to community well-being and solidarity. The ability of these alliances to create the kind of political power necessary to counter global forces depends on the capacity of movements to bring global issues down to the ground and root them in local concerns, creating a sense that "we are all in this together." This strategy not only creates common targets but also provides tangible solutions and manageable outcomes to problems—such as economic globalization, trade and investment law, and climate change—that are often seen as too complex and unwieldy when manifested at levels beyond the community.

Examining localized cross-movement coalition building in response to globalization is an important area for research on collective behavior. Many scholars have argued that cross-movement coalitions representing a broad spectrum of society are needed to counter the deleterious effects of globalization.[7] While there is considerable research on counter-globalization coalitions, much of the focus of this research is on the transnational level.[8] At the same time, some social movement scholars have argued that these coalitions are neither sustainable nor able to offer a concrete alternative to globalization because they lack the rooted networks, resources, and shared collective identities that characterize domestic movements.[9] Yet, my research demonstrates that these barriers do not necessarily hold true for movement alliances in local communities. Cross-movement coalitions at the local level have the capacity to foster long-term, sustainable movements because, unlike transnational coalitions, they have clear and tangible targets, pooled resources, rooted networks, and a shared sense of fate or collective identity in the face of external threats from increased globalization.

Toward a Theory of Localized Resistance to Globalization

The differences in frames, political contexts, and coalitions, despite shared understandings of water across the two cases, demonstrate the multiple processes of contention that shape anti–water privatization movements, beyond situated meanings of water. As the narratives from Stockton and Vancouver reveal, global processes—including multinational corporations, international financial and trade institutions, and global networks—interact with local political and organizational culture to shape the development, trajectories, and outcomes of local anti–water privatization movements.

Global processes affect local movements in three ways. First, globalization provides new opportunities and targets—in the form of global economic institutions and capital investment—for social movement actors to respond to and mobilize against. Recognition of global opportunity structures by social movement actors is important because local institutions are increasingly shaped by wider shifts in the global economy. Second, transnational contention fosters new networks that link transnational and local processes through the presence of *global connectors*—organizations and individuals with ties to both global and local movements. These network ties shape movement tactics drawn from broader anti-globalization movements. Third, global opportunities and transnational movement ties facilitate the formation of *global frames* that highlight the risks to local sovereignty from global financial institutions and flows of capital. Movements that recognize the global nature of local problems and strategically draw attention to the interplay between global and local processes bring together activists and political elites under a common sense of community well-being.

These findings advance the understanding of how "the global" is used by activists on the ground. For local movements implicated in transnational flows of capital and the power of international institutions, drawing attention to the shift in environmental risks—from the domestic to the transnational realm—legitimizes the claims of movement actors and facilitates new opportunities for shaping policies that resist commodification and corporate power. The construction of *global risk frames* through the presence of *global connectors* is of strategic importance for anti–water privatization movements because it draws attention away from the local conflict between activists and authorities to the international arena, creating a sense of *local solidarity* and a shared sense of fate in the face of global threats.

While Sidney Tarrow and Doug McAdam argue that the spread of contention from the local to the global introduces new conflicts and tensions to the initial movement by expanding the number of actors, frames, and targets involved, this is less likely to occur with the shift downward, from transnational to local movements, because the intimacy between local activists and organizations and a shared sense of community well-being works to mediate these differences.[10] The stories from the movements in Stockton and Vancouver reveal that local communities are far from powerless in the face of global institutions and capital. Indeed globalization has the potential to allow social movements to move beyond identity or class-based politics to a more broad-based and inclusive counter-hegemonic movement. What matters is the interplay between global processes and local movement-building conditions.

A Model for Other Movements

Taken together, each chapter in this book represents one part of a more comprehensive explanation of contention. The main goal is to understand what factors shape the outcomes of anti–water privatization movements. Yet the findings of the study can also be applied more generally, particularly for understanding the role of local social movements in the context of increased economic globalization. While much of the research on globalization reinforces the hegemonic power of global capital to transform societies—whether at the national, regional, or local levels—my research points to the power of communities to resist global economic hegemony and create alternatives to the deleterious consequences of neoliberal globalization.

The stories from the anti–water privatization movements in Stockton and Vancouver reveal important insights about the mechanisms that shape local movements in a global context and suggest some of the "best practices" that similar movements could potentially adopt in order to achieve their goals. The movement-building conditions demonstrated by the Vancouver case, including developing strong cross-movement coalitions, synthesizing global and local frames, and creating local solidarity in the face of international threats, may have important implications for other mobilizations at the local level that involve heterogeneous alliances. These movements include struggles against water privatization in the Global South and the Global North, and those emerging in response to the negative consequences of climate change and increasing inequality, including environmental-migrant or feminist-migrant coalitions. Some of key strategies for these movements to consider are:

• Identify alliances—including ties with political elites—at the local or regional level to inspire community solidarity and build cross-sectoral coalitions, which are more likely to emerge and strengthen in the face of external threats.

• Mobilize strong cross-movement coalitions that unite previously disconnected sectors, such as labor and environmental movements. Cross-sectoral coalitions are more successful with the presence of *frame bridgers*, organizations or individuals who help connect the concerns of diverse organizations and create common frames. Activists should identify groups that have ties to multiple and diverse movement sectors. Social justice groups are particularly important for bringing together labor and environmental organizations because they have ties to both movement sectors and are able to demonstrate the links between workplace issues and environmental justice.

• Reframe local problems in terms of global risks to local control over resources and services. Activists should actively identify *global connectors*, whose links to global and local movements enable them to draw on the frames, networks, and resources at the transnational level and help translate complex international issues into arguments that resonate at the local level.

• Target domestic political opportunity structures. While transnational frames, networks, and institutions provide *symbolic support* in the form of narratives and threats, ultimately it is most important for local activists to strategically connect global threats to local political process, and highlight the risks to local control and accountability, to bring together activists and elites under a share sense of community well-being.

What local counter-globalization movements need to remember is that successful challenges and alternatives to neoliberal globalization will not necessarily come from movements operating at the transnational level, but rather from locally situated counter-hegemonic movements that are connected globally but rooted in local communities. Unlike movements operating in transnational spaces, local resistance to globalization has the ability to generate long-term movements because of the presence of rooted networks, resources, and clear and tangible targets.

These local movements are connected to wider global movements and processes, allowing them to create new opportunities and draw on critical transnational networks and resources that help restructure power on the ground. With the rise of transnationalism and information technology, social movements are shaped by and dependent on local culture and values but are also connected globally through technologically mediated networks, which allow social movements to create new forms of resistance to global institutionalized power.[11] Without support from transnational networks, local movements that resist neoliberal globalization are "outmatched" by global corporate power and institutions.[12] As Mike O'Brian, global water campaigner and social justice activist, argues:

Communities have the power to fight back against globalization and corporate control. But it needs to work on two levels. Local communities need to be tapped into global networks as much as the international community needs the knowledge that comes from local communities. Locally, it makes the communities stronger to be plugged in to this global network in terms strengthening their claims and creating solidarity. But these connections also make mobilization much more effective internationally. To actually bring stories from those communities, and not just from developing communities, but from communities all over the world where these things have been happening, to directly intervene in discussions between governments, NGOs, and international institutions is very important. You need people speaking for the communities, for the people living in those places— just as you need to bring the global stories to the local communities in order to create more leverage for people facing these struggles.

O'Brian's comments reflect what many of the activists told me: that globalization is not inevitable, and that local democracy and autonomy can be a powerful force in the face of global flows of capital and regulatory institutions. Yet, for communities to successfully overcome the entrenchment of neoliberal globalization in local policy making and offer viable alternatives, global issues need to be brought down to the local level in ways that make sense on the ground.

Conclusion

Examining local-global linkages that shape movements at the community level is fundamental for advancing our understanding of the processes and mechanisms that shape contentious politics in a globalized world. While movements resisting neoliberal globalization have not yet succeeded in reversing the entrenchment of global economic hegemonic policies and practices, their continued presence, sustained mobilization, and alternative visions for more equitable and sustainable policies confront the inevitability of globalization and offer hope for the future. The challenge for these kinds of movements is to recognize the power of globalization and seize opportunities presented by transnational frames, networks, and institutions, while drawing on locally rooted networks and resources to build solidarity on the ground and create viable alternatives to global economic hegemony.

The forces that are impinging on local communities are increasingly global. This does not mean we should assume that the most important scale of resistance must then necessarily be at the global level. What happens at the local level matters for the outcomes of globalization. Indeed, some of the most consequential campaigns occur at the local level, where activists have successfully resisted neoliberal policies, from preventing water privatization to stopping the construction of Wal-Mart stores, to enacting living-wage ordinances. Understanding these local campaigns and their effects is critical for understanding social change in a global era.

Despite the eventual victories of the anti–water privatization movements in Stockton, Vancouver, and elsewhere, multinational water companies and global investment interests have not given up. As water becomes more scarce and its value continues to soar, private water firms have shifted attention from large urban centers—where they have been challenged by civil society and governments and where their contracts have often been rescinded for failure to deliver on their lofty promises—to smaller communities, in which governments and citizens are outmatched by the resources and expertise of these companies and their investors. Water still promises to be the new gold, and the fight continues between those who seek to outsource water resources and reap the profits and those who wish to protect water as a public resource. For those who oppose the privatization of water services and strive to build governance models based on ecological and social equity, success will not likely result from transnational mobilization and the targeting of international institutions, but rather from people coming together in their communities to take a stand against the growing commodification of the commons.

Notes

Introduction

1. Snitow and Kaufman 2007.
2. All of the names of individuals and organizations used in this study are pseudonyms to protect the identity of my respondents.
3. Public Citizen 2003.
4. Siders 2007.
5. Waters 2009.
6. Snitow and Kaufman 2007.
7. Ibid.
8. Tady 2007.
9. Simpson 2001.
10. Smith 2002.
11. Lees 2001.
12. Klein 2001.
13. Greater Vancouver Regional District 2001.
14. Bandy and Smith 2004, 1.
15. Keck and Sikkink 1998; Smith 2002; Bandy and Smith 2004; della Porta et al. 2006; Kay 2011.
16. Smith 2002.
17. della Porta et al. 2006.
18. Keck and Sikkink 1998; Smith and Johnston 2002; Ancelovici 2002; Kay 2011.
19. Keck and Sikkink 1998, 16.
20. Tarrow 2005; Tilly and Tarrow 2007.
21. Tarrow 2005, 42.
22. Keck and Sikkink 1998; Smith 2002; Fisher et al. 2005; della Porta et al. 2006; Kay 2011.

23. Ancelovici 2002; Josselin 2007; Leitner et al. 2007.

24. Burawoy et al. 2000; Josselin 2007; Hackworth 2007; Pellow 2007; Evans 2008; Chun 2009.

25. Gould, Schnaiberg, and Weinberg 1996, 4.

26. Roberts and Toffolon-Weiss 2001, 191.

27. Roberts and Toffolon-Weiss 2001.

28. Blatter, Ingram, and Doughman 2001.

29. See O'Brian 1992.

30. Blatter, Ingram, and Doughman 2001.

31. Andrews 2004; Andrews and Edwards 2005.

32. Leitner et al. 2007; Sassen 2000.

33. One notable exception is Karen Bakker's research on water privatization in England and Wales (Bakker 2004).

34. Boelens, Getches, and Guevera-Gil 2010; Bakker 2010; Morgan 2011.

35. In Vancouver, these groups included City Green, a local environmental organization, the British Columbia Public Service Employees Union (BCPSEU), and the Citizens Action League, an organization focused on social, economic, and environmental justice. In Stockton, the groups included the Association of Democratic Voters, the Citizens Environmental Alliance, and the Utility Workers Local 5 Union.

36. Cress and Snow 2000; Voss and Sherman 2000; Tarrow 2010. A comparative, qualitative approach is the best way to uncover the complex processes of social movements (Voss and Sherman 2000; McAdam 2003; Lopez 2004). Qualitative methods move beyond external causes and allow for an examination of the processes through which activists work, including the interactions of identities, ideologies, and discursive strategies (Lopez 2004; Broadbent 2003; Mische 2008). Tarrow (2010) argues that unlike large-N quantitative studies, paired comparisons offer the "intimacy of analysis" that allows for an in-depth examination of "how common mechanisms are influenced by the particular features of each case," while at the same time enabling the development of hypotheses not possible in single-case studies (2010, 245–246).

37. To identify and recruit potential respondents for the study, I began by researching archival documents, including media stories, minutes of public meetings, and organizational Web sites to identify key leaders and social movement organizations involved in the anti–water privatization movements in Stockton and Vancouver. Each potential respondent then received a letter of introduction with a request for a face-to-face interview. I used "snowball" sampling to secure interviews with other members of the movements by asking these key leaders to recommend other potential respondents for interviews.

38. Tarrow 2010.

39. Shiva 2002, x.

Chapter 1

1. Rothfeder 2001; Conca 2005; Whiteley, Ingram, and Perry 2008.

2. Catley-Carlson 2003.

3. Conca 2005; Bakker 2010.

4. Bakker 2010.

5. For example, in the "Dublin Statement on Water and Sustainable Development," the United Nations recognized the fundamental right of all humans to have access to an affordable and adequate supply of clean water and sanitation (United Nations 1992). In the last two decades, several countries have enshrined the human right to drinking water into their constitutions, including South Africa and Bolivia (Ingram, Whiteley, and Perry 2008). At the same time, a growing global water movement continues to advocate for water to be declared a human right by governments around the world (Bakker 2010).

6. Whiteley, Ingram and Perry 2008; Bakker 2010.

7. Mullin 2009; Bakker 2010; Morgan 2011.

8. Bauer 2004; Mullin 2009; Bakker 2010.

9. Bauer 2004; Bakker 2010. These conferences include the International Conference on Water and Environment, held in Dublin in 1992, the United Nations Conference on Environment and Development (the Earth Summit in Rio De Janeiro) in 1992 and the 2nd World Water Forum in The Hague in 2000 (Bauer 2004).

10. Bauer 2004.

11. United Nations/World Water Assessment Programme 2003.

12. The "Dublin Statement on Water and Sustainable Development," also known as the Dublin Principles, adopted four main principles regarding water. (1) "Fresh water is a finite and vulnerable resource, essential to sustain life, development and the environment"; (2) "Water development and management should be based on a participatory approach, involving users, planners and policy-makers at all levels"; (3) "Women play a central part in the provision, management and safeguarding of water"; (4) "Water has an economic value in all its competing uses and should be recognized as an economic good" (United Nations 1992).

13. Wilder 2008; Bakker 2010; Morgan 2011.

14. Mullin 2009; Bakker 2010.

15. Mullin 2009; Morgan 2011.

16. Conca 2006.

17. Mullin 2009; Bakker 2010.

18. Bauer 2004; Wilder 2008; Bakker 2010.

19. Mullin 2009.

20. Mullin 2009, 3.

21. Hackworth 2007; Morgan 2011.

22. Morgan 2011.

23. Morgan 2011, 6.

24. Barlow and Clarke 2002; Bauer 2004; Goldman 2007; Bakker 2010.

25. Barlow and Clarke 2002.

26. Goldman 2007; Bakker 2010.

27. Barlow and Clarke 2002; Schoenberger 2003.

28. Standing 2002; Tickell and Peck 2003; Harvey 2005; Hackworth 2007; Bakker 2010.

29. Harvey 2005; Tickell and Peck 2003.

30. Standing 2002; Tickell and Peck 2003; Harvey 2005.

31. Stiglitz 2002; Goldman 2007; Hackworth 2007; Morgan 2011. Morgan (2011, 100) argues that international trade and investment agreements cast a "powerful political shadow" over the global water policy field and have growing influence over domestic decision making and environmental regulation. She argues that these agreements "provide for any disputes to be resolved by arbitration processes that tend to exclude broader public interest issues or even to view national regulation as potentially expropriatory (Morgan 2011, 100)."

32. A 2006 report by the Canadian Imperial Bank of Commerce (CIBC) revealed that the World Bank estimates a doubling of water privatization by 2011 across the globe, encompassing almost 40 percent of the market (Tal 2006).

33. Peck and Tickell 2002.

34. Clarke 2003; Goldman 2007.

35. Rothfeder 2001; Goldman 2007.

36. Low and Gleeson 1998.

37. Goldman 2007. Michael Goldman refers to this phenomena as "green neoliberalism" and argues that the "pro-poor" policy promoted by the World Bank and other global financial institutions is driven by three factors: the emergence of an international network of elite policy makers, who are increasingly seen as the new water experts, the linking of development loans to private sector investment in water services, and the implementation of what he calls "green-neoliberal" loan conditions for countries wishing to access foreign capital (2007, 790).

38. Bakker 2010, 43.

39. Rothfeder 2001; Bakker 2010. Rothfeder (2001) argues that the track record of public sector water utilities and governments around the world has been dismal, leading to an enormous financial shortfall for fixing the problems caused by poor governance and management. He contends that the United States alone needs more than $23 billion per year just to prevent its water systems from deteriorating further and argues that in some Eastern European countries "the amount of money required to build and fix water systems is more than the gross national product" (106). Karen Bakker (2010) argues that governance failure has occurred under publicly run water systems—what she calls "state failure"—while, at the same time, private sector models, for the most part, have failed to correct the problems related to public sector water management. She argues that regard-

less of the management model, strong public sector oversight is needed to ensure proper regulation, fair distribution, and good governance of water resources.

40. Bakker 2010.

41. Ibid.

42. Goldman 2007.

43. Snitow and Kaufman 2007; Bakker 2010.

44. Rothfeder 2001.

45. Rothfeder 2001; Olivera and Lewis 2004.

46. Rothfeder 2001; Bakker 2004.

47. Wilder 2008; Arnold 2008.

48. Bauer 2004; Bakker 2010.

49. Shiva 2002; Goldman 2007.

50. Bauer 2004; Wilder 2008.

51. Boelens, Getches, and Guevara-Gil 2010.

52. Bakker 2004, 2010; Conca 2005; Castro 2007; Goldman 2007.

53. Bauer 2004; Ingram, Whiteley, and Perry 2008; Morgan 2011.

54. Ingram, Whiteley, and Perry 2008; Arnold 2008; Wilder 2008; Bakker 2010; Boelens, David Getches, and Guevara-Gil 2010.

55. Ingram, Whiteley, and Perry 2008, 2.

56. Arnold 2008. Thomas Clay Arnold argues that "water is not simply a natural resource for ever-more efficient or profitable consumption, the underlying assumption of many market-related policy proposals. Water is much more. It is also a social good, the basis for, among other things, uniquely valued identities, relationships, and civic associations. Policies that downplay or ignore one or more of the many uses, meanings, and values of water increase the chances for political conflict and defeat (2008, 55)."

57. A critique of commodification does not mean that the value of water is not important. Many water economists, for example, argue that water has historically been considered an economic good due to its scarcity. Yet the status of water as an economic good does not mean it must be treated as a commercial commodity. What is important is the recognition of water as a valuable and scarce resource, and thus trade-offs are required to ensure it is not wasted (Bauer 2004).

58. Bauer 2004; Ingram, Whiteley, and Perry 2008; Bakker 2010.

59. Blatter, Ingram, and Doughman 2001, 20.

60. Blatter, Ingram, and Doughman 2001, 14.

61. Bauer 2004; Whiteley, Ingram, and Perry 2008; Wilder 2008; Bakker 2010; Morgan 2011.

62. Wilder 2008; Bakker 2010; Morgan 2011.

63. Bakker 2010.

64. Wilder 2008.

65. Wilder 2008, 102.

66. Bauer 2004; Wilder 2008; Bakker 2010.

67. Ingram, Whiteley, and Perry 2008; Arnold 2008. Arnold argues that policies should be based on what he calls complex equity, which is "a form of distributive justice, a regulative ideal for goods distinguished by their multiple meanings and values. According to the principle of complex equity, justice rests on managing complex social goods in light of their multiple meanings and values, and on rendering those meanings and values their due" (2008, 52).

68. Wilder 2008; Boelens, Getches, and Guevara-Gil 2010; Bakker 2010; Morgan 2011.

69. Blatter, Ingram, and Doughman 2001; Wilder 2008.

70. Wilder 2008, 107.

71. Boelens, Getches, and Guevara-Gil 2010.

72. Wilder 2008. Wilder argues that for water governance policies to be effective, they must bring together different dimensions of equity, including political equity —"the institutionalization of local participation in water policy-making as well as the quality of the participatory mechanisms," and economic equity—"the availability, accessibility and affordability for those who are less well off (2008, 95)."

73. Bakker 2010, 202.

74. Ibid, 194.

75. Ibid. Bakker argues that private sector involvement in water management must be accompanied by strong regulatory oversight by the public sector to ensure that equity issues are an important part of policy decisions. Other scholars have argued that what matters for effective water governance is not whether the system is public or private, but rather that the political context of water is considered when implementing policies. For example, Marco Schouten and Klaas Schwartz argue that water should be treated as a political good and claim that "sound and sustainable investment in water services can only be achieved by taking into account the existing political environment in which those water services are to be delivered" (Schouten and Schwartz 2006, 1). What matters therefore is the correct balance between private sector investment and government oversight.

76. Bakker 2010; Morgan 2011.

77. Shiva 2002; Bakker 2010.

78. Bakker 2010, 137.

79. Barlow and Clarke 2002; Shiva 2004; Terhorst 2008; Bakker 2010; Morgan 2011.

80. Rothfeder 2001.

81. Olivera and Lewis 2004

82. Rothfeder 2001; Baer 2008.

83. Olivera and Lewis 2004.

84. Schouten and Schwartz 2006; Baer 2008.

85. Evans 2000; Munck 2002; Peck and Tickell 2002; Dryzek 2006.

Chapter 2

1. McAdam, Tarrow, and Tilly 2001; Meyer 2004; Tilly and Tarrow 2007.

2. Beck 1999; Pellow 2007; Gould, Pellow, and Schnaiberg 2008.

3. McNaghten and Urry 1998, Satterfield 2002.

4. Bauer 2004, Ingram, and Perry 2008, Bakker 2010, Morgan 2011.

5. Urry 2000, 2003; Castells 2003; Presas and Mol 2006.

6. Urry 2000, 211. Urry distinguishes between traditional sociological concepts of vertical mobility, including educational, income, occupational, and social mobility and what he refers to as new horizontal mobilities that encompass social and geographical spaces, including flows of people, waste and pollution, ideas, and objects across social and geographic boundaries, in the context of a globalizing world.

7. Goodwin and Jasper 1999.

8. Gamson and Meyer 1996; Poletta 1999; Benford and Snow 2000; McAdam, Tarrow and Tilly 2001; Tarrow and Tilly 2007; Mische 2008.

9. Diani 1996; Andrews 2004; Pellow 2007; Staggenborg and Lecomte 2009.

10. McAdam, Tarrow, and Tilly 2001; Tarrow and Tilly 2007; McAdam, Tilly, and Tarrow 2008.

11. McAdam, Tarrow, and Tilly 2001, 25–26.

12. McAdam, Tarrow, and Tilly (2001) argue that social movement scholars should adopt an interactive and dynamic model for explaining contentious politics, based on identifying mechanisms and processes and how they combine and recur to shape episodes of contention. They claim that "by understanding which mechanisms and processes put an episode of contention in motion and where they take it, we can better understand why some episodes are brief while others are protracted, why some end in demobilization while others expand into revolution, and why some produce fundamental shifts in alignments and political culture why others leave behind nothing but a residue of bitter memories" (67).

13. Burnstein, Einwohner, and Hollander 1995; Andrews 1997; Tarrow and Tilly 2007; Armstrong and Bernstein 2008.

14. McAdam, Tarrow, and Tilly 2001; Tarrow and Tilly 2007.

15. Eyerman and Jamison 1991; Goodwin and Jasper 2004; Oliver and Johnston 2005.

16. Oliver and Johnston 2005; Schurman and Munro 2006.

17. Oliver and Johnston 2005.

18. Goodwin and Jasper 2004; Oliver and Johnston 2005; Schurman and Munro 2006.

19. Schurman and Munro 2006, 4.

20. Other scholars argue that individual identity and ideology not only transform movements themselves through the creation of collective identity and frames, but also challenge the dominant cultural codes and societal norms, thus creating the

potential for social change (Melucci 1988; Eyerman and Jamison 1991). Eyerman and Jamison argue that the ideological dimension of social movements—what they call "cognitive praxis"—plays a fundamental role in creating new knowledge and understandings of the world and thus should be a key area of investigation in social movement research (1991, 48).

21. McNaghten and Urry 1998; Satterfield 2002; Loftus and Lumsden 2008.

22. Blatter, Ingram, and Doughman 2001; Mullin 2009; Bakker 2010.

23. Ingram 1990; Blatter and Ingram 2001; Whiteley, Ingram, and Perry 2008; Mullin 2009; Morgan 2011.

24. Espeland 1998; Blatter, Ingram, and Doughman 2001; Morgan 2011.

25. Bakker 2010.

26. McNaghten and Urry 1996; Beck 1999; Blatter, Ingram, and Doughman 2001; Dauvergne 2008.

27. Ingelhart 2008.

28. Beck 1999; Blatter, Ingram, and Doughman 2001; Bakker 2010.

29. Catton and Dunlap (1978) and Dunlap and Catton (1994) refer to this shift in values as the transformation from the "human exemptionalist paradigm" (HEP), in which humans are considered superior to the environment (this was the dominant worldview from the Industrial Revolution to the mid–twentieth century), and the "new environmental paradigm" (NEP), which stresses the interdependence of humans and the environment and the need to recognize the ecological impact of social processes. They argue that the NEP is increasingly replacing HEP as the dominant worldview.

30. Dunlap and Catton 1994; McNaghten and Urry 1998; Hannigan 1995; Urry 2000; Dunlap and Michelson 2002.

31. Satterfield 2002.

32. Espeland 1998; Satterfield 2002; Loftus and Lumsden 2008.

33. Espeland 1998; Bauer 2004; Mullin 2009; Bakker 2010.

34. Whiteley, Ingram, and Perry 2008; Mullin 2009; Bakker 2010.

35. McNaghten and Urry 1998.

36. Oliver and Johnston 2005; Schurman and Munro 2006.

37. Snow et al. 1986; Goodwin and Jasper 1999; Polletta 1999; Benford and Snow 2000.

38. Goffman 1974; Snow et al. 1986; McAdam, McCarthy, and Zald 1996; Benford and Snow 2000; Johnston and Noakes 2005.

39. Benford 1993; Carroll and Ratner 1997.

40. Diani 1996; Benford 1997; Ferree et al. 2002; Johnston and Noakes 2005.

41. Reese and Newcombe 2003.

42. Beck 1999; Dauvergne 2008.

43. Dauvergne calls these consequences "ecological shadows of consumption" and argues that "the processes channeling these consequences occur inside a glob-

al system so complex, so chaotic, that tracing the pathways of cause and effect is beyond the traditional tools of policy makers or scientists (2008, 16)."

44. Beck 1999; Dauvergne 2008.

45. Bakker 2010; Morgan 2011.

46. Hamel et al. 2001; Smith 2002; Ancelovici 2002; Conway 2004.

47. Ancelovici 2002, 428.

48. McAdam 1996, 1999; Jenkins and Klandermans 1995; Tarrow 1998; Meyer 2004.

49. McAdam 1996, 1999; Tarrow 1998; Meyer 2004.

50. Tarrow 1998; McAdam, Tarrow, and Tilly 2001; Meyer 2004, Tilly and Tarrow 2007.

51. Ayres 2002; Pellow 2007; Kay 2011.

52. Joppke 1993; Goodwin 2001; Dryzek et al. 2003.

53. Goodwin and Jasper 1999; Pellow 2001; Kay 2005.

54. Ancelovici 2002; Meyer 2004; Pellow 2007; Schurman and Munro 2009.

55. Keck and Sikkink 1998; Smith 2002; Khagram, Riker and Sikkink 2002; della Porta et al. 2006; Pellow 2007; Kay 2011.

56. Smith 2002.

57. Smith 2002; della Porta et al. 2006.

58. Smith and Johnston 2002; Ancelovici 2002; Kay 2005; Evans 2008.

59. Tarrow and McAdam 2003.

60. Chun 2009.

61. Guidry, Kennedy, and Zald 2000; Ancelovici 2002; Josselin 2007.

62. Sassen 2000.

63. Hess 2009.

64. Roberts and Toffolon-Weiss 2001; Juris 2004; Dupuis and Block 2008; Hess 2009.

65. Dupuis and Goodman 2005, Hess 2009.

66. Dupuis and Goodman 2005, 364.

67. Gould, Schnaiberg, and Weinberg 1996.

68. Bakker 2010, Morgan 2011.

69. Magnusson 2005.

70. Peck and Tickell 2002; Meyer 2003; Castells 2003; Dryzek 2006.

71. Meyer 2003.

72. See Evans 1995; Peck and Tickell 2002; Dryzek 2006.

73. Pellow 2007; Walker, Martin, and McCarthy 2008; Evans 2008; Schurman and Munro 2009.

74. Pellow 2007, 62.

75. Andrews 2004; Andrews and Edwards 2005.

76. Urry (2000) contends that "[n]either the global nor the local can exist without the other. They develop in a symbiotic, irreversible and unstable set of relationships, in which each gets transformed through billions of iterations worldwide." This complex relationship between the global and the local can result in what he calls "globalisation-deepening-localisation" (210).

77. Diani 1995; Roth 2003; Baldassari and Diani 2007; Mische 2008.

78. Keck and Sikkink 1998; Khagram 2004.

79. McAdam, Tarrow, and Tilly 2001; Diani 2003; Diani and Bison 2004; Evans and Kay 2008; Baldassari and Diani 2007; Mische 2008.

80. Evans and Kay 2008.

81. Mische 2008.

82. Rose 2000; Roth 2003; Pellow 2007; Mische 2008.

83. Rose 2000.

84. Obach 2004.

85. A notable exception is the research presented in the book *How Social Movements Matter* (Guigni, McAdam, and Tilly 1999), which examines social movement outcomes, including policy changes, institutional impact and public discourse and highlights the need to move beyond analysis of movement emergence and development to examine the long-term social, cultural, and political outcomes of social movements.

86. Diani 1995; Tindall 2004; Lopez 2004.

87. Voss and Sherman 2000; Lopez 2004; Chun 2009.

88. Rose 2000; Obach 2004; Pellow 2007; Evans 2008; Brulle and Jenkins 2008.

89. Blatter, Ingram, and Doughman 2001.

90. Beck 1999, 14.

91. Beck 1999; Blatter, Ingram, and Doughman 2001.

92. Juris 2004, 346.

93. Castells 2003; Juris 2004; Evans 2008.

94. Keck and Sikkink 1998; Khagram 2004.

95. Khagram 2004.

96. See, for example, Keck and Sikkink 1998; Smith 2002; and Kay 2005.

97. Diani and Bison 2004.

98. Rose 2000; Lopez 2004; Pellow 2007; Evans 2008.

99. Sassen 2000; Peck and Tickell 2002; Evans 2000; Dryzek 2006.

Chapter 3

1. Gottlieb 1989; Espeland 1998; Ingram, Whiteley, and Perry 2008; Boelens, Getches, and Guevara-Gil 2010.

2. See Beck 1992; 1999.

3. Environmental meanings are mediated through the interplay of our senses with geographical location (Urry 2000; Greider and Garkovich 1994; McNaghten and Urry 1998). For example, in their study of how people in Lancashire County in the United Kingdom perceive and understand environmental and sustainability issues in their daily lives, McNaghten et al. (1995) found that individuals rely on direct sensory experience and observation, rather than scientific evidence or media reports, to make sense of environmental risks and changes in their communities.

4. Earthjustice 2008.

5. McNaghten and Urry 1998.

6. Despite a lower support for public investment in services by Americans as compared to their European counterparts (Alesina and Glaeser 2004), recent polling across the United States demonstrates that a vast majority (84 percent) of Americans support public investment in water treatment systems and government investment in infrastructure in general (Luntz 2009). Further, 83 percent of Americans support legislation to create a national clean water trust fund to invest in water infrastructure maintenance and upgrades (Kranz 2004).

7. Anti–water privatization activists have deployed the frame of the human right to water to counter what they see as a growing trend toward the commodification of water and its treatment as an economic good (Bakker 2010). Activists claim when water is considered a human right, it becomes incompatible with privatization, as pricing water will necessarily prevent access for many people in poor or disadvantaged communities due to its unaffordability (Bauer 2004; Barlow and Clarke 2002). Some water scholars agree with this claim and suggest that treating water as a human right—under an international human rights framework—will overcome the problem of access and affordability by setting a minimum required daily standard of water in order to meet people's basic daily needs and guarantee that vulnerable populations have access to water (Gleick 1996; Bluemel 2004). At the same time, other water scholars dispute the viability of the human rights framework and argue that the debate should focus on issues of equity rather than on public versus private ownership and water rights. Geographer Karen Bakker, for example, argues that defining water as a human right is problematic in several ways. First, the term itself is difficult to define and the complexity of the international human rights legal arena makes it challenging to implement. Second, Bakker points to the anthropocentric nature of "human" rights and claims that this argument ignores or downplays the rights of nonhumans, including animals and ecosystems. Third, the concept of human rights seeks to apply an individualistic, rationalistic, Western legal paradigm to water governance that does not take into account different cultural, religious, or spiritual understandings of water in communities around the world. Bakker's main concern is that under a human rights framework, water rights "are inevitably transformed into bureaucratic technical formulations (such as a "minimum" amount of water per person per day that satisfies the human right to drinking water) that evade broader questions of social justice" (2010, 159). Bakker, along with other prominent water scholars, including Helen Ingram, John Whiteley, and Richard Perry, contend that an equity

framework is more conducive to ensuring distributive social and environmental justice because it compels policy makers to take into account and address issues of political, social, and economic inequality (Bakker 2010; Whiteley, Ingram, and Perry 2008).

8. In 1999, the Metro Vancouver Board of Directors adopted a no-logging policy for the North Shore watersheds that supply water to the region. This new policy was the result of a decade-long campaign by environmental groups to stop industrial logging in the watersheds because of threats to water quality and soil erosion (Society Promoting Environmental Conservation 1999).

9. Concerns over the effect of outsourcing on the quality of the water supply were also shaped by previous episodes of water-borne illness that had recently occurred across Canada. In May 2000, thousands of people became ill and seven people died as a result of drinking water contaminated with E. coli bacteria in Walkerton, Ontario (Perkel 2002; Ali 2004). Many people have argued that the privatization of water-testing laboratories, combined with the neoliberal policies of the provincial government at the time, led to mismanagement of water quality reporting and a loosening of environmental and health regulations (Perkel 2002; Prudham 2003; Ali 2004).

10. Pellow (2007) argues that environmental justice movements in the United States that emphasize broader human rights concerns are more successful than those that focus only on issues of race and class. He contends that connecting environmental justice concerns with human rights provides the way forward for such movements to be successful in their claims because human rights is a globally recognized and legally enshrined concept and thus resonates widely with political opportunity structures, both locally and internationally.

11. Espeland (1998) supports Mark Spencer's sense that the emotional dimension of water drives the passion in local water politics, channeling individual anger and deep-seated feelings about local water issues into political action.

12. Bakker's (2007) research on municipal water privatization points to the strong sense of public ownership that people express when discussing water. She contends that movements that frame anti–water privatization arguments around the concept of the commons are more successful than those that rely on other arguments, including water as a human right, because they are able to counter the notion of water as a commodity and offer alternative models for resource management based on the idea of a shared commons.

13. Ingram, Whiteley, and Perry 2008; Boelens, Getches, and Guevara-Gil 2010.

14. Swidler 1986.

15. McNaghten and Urry 1998; Espeland 1998; Satterfield 2002; Bakker 2010.

16. Weber (2003 [1905]) argued that social behavior in modern society—characterized by a growth in bureaucratic institutions and the dominance of the capitalist system—is increasingly shaped by instrumental rationality rather than the kinship values reflective of preindustrial society. Wendy Espeland's (1998) research on the conflict over dam construction in Arizona, in particular, challenges the notion of rational choice and demonstrates that people's "rationalities" and conceptions of water are far from instrumental, but rather are shaped by identity, power, and emotion.

17. Geographer Eric Swyngedouw (2004) argues that water resources and water governance structures are embedded in what he terms the "hydrosocial flows" of water; the multiple sets of power relations—social, cultural, economic, and political—that shape who has access to and control over water. He contends that water governance is as much about issues of equity as it is about environmental protection and conservation.

18. Bauer 2004; Bakker 2010; Boelens, Getches, and Guevara-Gil 2010; Morgan 2011.

Chapter 4

1. Matsusaka 2004; Gerber and Phillips 2005.

2. Gould, Schnaiberg, and Weinberg 1996; Tarrow and McAdam 2005; Ancelovici 2002; Pellow 2007.

3. Olivera and Lewis 2004.

4. While there are clear examples of negative outcomes of water service outsourcing, including the failure to upgrade infrastructure and equity in terms of access and affordability (see Rothfeder 2001; Bakker 2007; and Wilder 2008), one should not necessarily assume that private sector water service delivery is incompatible with ensuring equity in access and sound stewardship of water resources. Some scholars, including Karen Bakker and Marco Schouten and Klass Schwartz argue that private sector participation in water governance can generate positive outcomes, in terms of generating investment for infrastructure upgrades and ensuring efficiencies in delivery and conservation. These scholars contend that problems are related in governance failure and this can occur under both private and public sector management. What matters, they argue, is governance oversight through regulation and policies that ensure both stewardship and distributive justice (Schouten and Schwartz 2006; Bakker 2010). Bakker, for example, claims that good water governance needs to include an ecological dimension to guarantee the stewardship of water resources, while recognizing the multi-scalar nature of water, both geographically and politically (Bakker 2010). Schouten and Schwartz maintain that companies need to pay attention to the political context of water governance—including the multiple and competing demands on water resources as well as the wider social and economic context—to ensure best practices in water governance and service delivery oversight (Schouten and Schwarz 2006).

5. While movement organizers in Vancouver focused a large part of their attention on the struggle over water in Cochabamba, Bolivia (both because of the dramatic outcomes of water privatization and the fact that one of the main movement leaders from Cochabamba had recently visited Vancouver), they also drew attention to other communities around the world that had outsourced their water services. During the public consultation process, anti–water privatization activists distributed informational pamphlets, created by the BCPSEU, that highlighted the negative outcomes of water privatization in other communities, including Cochabamba, Bolivia; Orange Farm, South Africa; Hamilton, Ontario; Grenoble,

France; and Atlanta, Georgia. Although these cases are not necessarily representative of the outcomes of municipal water privatization in communities around the world, they were used strategically by movement actors in Vancouver as symbolic leverage to mobilize people into action and sway the opinions of policy makers.

6. This critique of the global capitalist economy and its need for continuous economic expansion reflect what Schnaiberg and Gould (2000) label the "treadmill of production" that results in the enduring conflict between the economy and the environment (see also Gould, Pellow, and Schnaiberg 2008).

7. Castells 2007; Pellow 2007; Evans 2008.

8. Ancelovici 2002; Olesen 2005; della Porta et al. 2006.

9. NAFTA's Chapter 11 investor clause is a dispute mechanism that allows investors in any of the three signatory countries—Canada, the United States, and Mexico—to sue the federal government of one of the other two signatory countries to compensate for lost profits when legislation by any level of government is implemented that would impinge on the investor's future profit-making ability. Chapter 11 provisions have already been utilized by Sun Belt Water Inc., a California-based company, to sue the federal government of Canada for lost profits resulting from legislation in British Columbia banning bulk water exports (Gleick 2002).

10. Citizens Action League 2001.

11. Canada has a long history of concerns about sovereignty especially in reference to increased economic integration with the United States (Bolt 1999; Adams 2003; Finlay 2004). While differences between the frames deployed by the movements in Vancouver and Stockton are perhaps partially explained by overarching value differences between Canadians and Americans (for example, in his book *Fire and Ice* [2003], Michael Adams argues that fundamental value differences distinguish Canadians from their American counterparts—including perceptions of the role of government and religion—leading him to conclude that Canada and the United States are socioculturally distinct), the responses from activists in Vancouver reveal understandings beyond simplistic anti-American, anti-corporate biases, and reflect a sophisticated understanding of the power of global capital to encroach on local sovereignty and the risks posed by the Canada's legal obligations under international trade and investment treaties on municipal decision making and regulatory power. These understandings are linked to the presence of *global connectors* that used transnational ties, resources, and flows of information to locate the problem of local water privatization in the broader context of neoliberal globalization.

12. Beck 1999; Urry 2003.

13. Epp 1998; McCaan 2004; Bernstein, Marshall, and Barclay 2009.

14. Meyer and Staggenborg 1996.

15. Sassen 2000; Peck and Tickell 2002; Hackworth 2007; Evans 2008.

16. Gamson 1990; Benford 1997; Diani 1996; Cress and Snow 2000.

17. Tarrow 2005.

Chapter 5

1. Peck and Tickell 2002; Meyer 2004; Dryzek 2006; Pellow 2007; Evans 2008.

2. The city of Stockton has a population of 291,707. In 2010 the average household income was $52, 646 ("State & County QuickFacts—Stockton, California," U.S. Census Bureau, http://quickfacts.census.gov/qfd/states/06/0675000.html). Stockton is a charter city that is governed by an elected city council that appoints and oversees fifteen departments, including the city manager, city clerk, city auditor, and city attorney's office and service provision departments such as fire, water, and sewage and library services, parks and recreation, police, and public works. The economy is dominated by the agriculture and manufacturing sectors, including warehousing and distribution ("Stockton, California," Open World Leadership Center, http://www.openworld.gov/hosts/city.php?id=1108&lang=1). The population of the Metro Vancouver region is 2,097,965 (the population of the city of Vancouver is 571,600). In 2006 the average household income was $73,258. The major economic sectors are services, trade, and tourism. The Metro Vancouver region is composed of twenty-two municipalities and one electoral district. It has three main governing roles in the region; the delivery of service—including the provision of drinking water, the treatment of sewage, and the management of solid waste—overseeing planning and providing political leadership. The Metro Vancouver Board is made up of thirty-seven directors, elected representatives from the member municipalities who are appointed by their respective city councils ("Who Is Metro Vancouver," Metro Vancouver, http://www.metrovancouver.org/about).

3. Mullin 2009; Bakker 2010.

4. Rothfeder 2001; Mullin 2009.

5. Rothfeder 2001; Bauer 2001, Mullin 2009; Bakker 2010

6. Rothfeder 2001, 103.

7. Snitow and Kaufman 2007.

8. Hackworth 2007; Thomas 2008; Mullin 2009.

9. Moulton and Anheim 2000.

10. Snitow and Kaufman 2007.

11. Joppke 1993; Barclay, Burstein, and Marshall 2009.

12. Gerber 1999; Matsusaka 2004; Gerber and Phillips 2005; Martin 2009. Of course, critics of the initiative process contend that they are open to manipulation by powerful interest groups and often used to advance legislation that disadvantages vulnerable populations, including racial and ethnic minorities (HoSang 2010), although this claim has been disputed by some scholars. For example Elizabeth Gerber (1999) and John G. Matsusaka (2004) argue that most initiatives in the United States over the last thirty years have been supported by the majority of voters and do not disproportionately disadvantage minority or low income populations. At the state level, the ballot initiative process has often been used in an attempt to dismantle progressive policies and public programs that largely benefit low income people, communities of color, and immigrant populations (HoSang

2010). In recent decades, ballot initiatives have been launched to deny public benefits to undocumented immigrants, end affirmative action policies, overturn gay marriage legislation, or enact welfare reform (Garcia 1995; Bosco 1994; Miller 1999). Ballot measures can also create barriers to state legislatures seeking to raise revenue for new programs, leaving governments unable to fund current or future services (Martin 2009).

13. Gordon 2004.

14. Ibid.

15. Hilson (2002) argues that the role of law and the courts has been downplayed in the literature on social movements and contentious politics and suggests that more attention should be paid to legal opportunity structures and how they influence movement outcomes. Yet the findings of my research also reveal that relying too heavily on legal tactics can be detrimental to social movement success as organizations shift attention and resources away from mobilizing broad-based support and targeted action designed to bring about policy change. In some movements, legal tactics combined with organized collective action is a strategy more likely to succeed than a reliance on lawsuits alone.

16. Gamson 1990; Andrews 2004; Amenta and Caren 2004.

17. Clarke 2003.

18. Barlow and Clarke 2002; Bakker 2007.

19. Carroll 1993.

20. Clarke 2003; Bakker 2007.

21. Keil 2002.

22. See Ancelovici 2002; Staggenborg and Lecomte 2009.

23. Pellow (2007) calls this the political economic process perspective.

24. Many scholars have argued that there are important value and cultural differences between Canada and the United States that have led to the diverse economic and social trajectories of the two countries. For example, while the Canadian government is more interventionist in terms of social and labor market policies, the United States relies more on unrestricted market forces (Lipset 1990; Thomas 2008). These values have resulted in a less interventionist government than its counterpart in Canada. While it would be reasonable to assume that these national institutional and value differences might explain the divergent trajectories and outcomes between the anti–water privatization movements in Vancouver and Stockton, particularly the willingness of the regional government in Vancouver to protect public sector services, some evidence suggests that other factors are more important for explaining differences between these cases, including the deployment of global risk frames and targeting of international institutions. For example, while the movement in Vancouver was successful at preventing water privatization, other communities in Canada, including Hamilton, Ontario, and Moncton, New Brunswick, have failed to block the outsourcing of public water services through mobilization, despite the presence of anti–water privatization movements (Barlow and Clarke 2002). At the same time, social movement mobilization has successfully prevented the privatization of municipal water services in several U.S. cities, including New Orleans, Louisiana, and Felton, California (Sni-

tow and Kaufman 2007). In both countries, decisions about the regulation and control of water resources are made at the municipal rather than national level. And while Canada-U.S. value differences are not insignificant, when it comes to values around water, Canadians and Americans share very similar beliefs in the importance of public sector protection, regulation, and delivery of water services (Luntz Research Companies 2005; Nanos Research 2008).

25. Pellow 2007; Schurman and Munro 2009.

26. Sassen 2000; Hackworth 2007.

27. Bulkeley and Betsill 2003, 2.

28. Hamel et al. 2000.

Chapter 6

1. Shiva 2002; Barlow and Clarke 2002; Bakker 2010; Morgan 2011.

2. Rose 2000; Lopez 2004; Obach 2004a.

3. Rose 2000; Obach 2004a; Evans 2008.

4. Conca 2005; Bakker 2010.

5. Rose 2000; Gottlieb 2002; Lopez 2004; Obach 2004a.

6. Burawoy (2010) argues that the commodification of nature, including water, poses not only a threat to the environment, but also threatens labor through downward pressure on wages. This dual threat could potentially act to build solidarity between labor unions and other social movement organizations that seek to protect nature from commodification.

7. Staggenborg 1986; Van Dyke 2003.

8. Obach 2004a.

9. Wilson 1998; Samuel 2009.

10. Gottlieb 2002; Obach 2004a.

11. Rose 2000; Obach 2004a.

12. Obach 2004b.

13. Dowie 1996.

14. Gottlieb 2002.

15. Dowie 1996; Gottlieb 2002.

16. Obach 2004a; Gould, Pellow, and Schnaiberg 2008.

17. Clawson 2003; Lopez 2004; Chun 2009.

18. Diani and Rambaldo 2007.

19. In 2008, Stockton had the highest rate of foreclosures in the United States with one foreclosure for every twenty-seven households, representing a 256-percent increase compared to 2007 ("RealtyTrac" 2008).

20. Research on alliances between labor and other social movements demonstrates that labor often provides resources and support for other movements, including environmental and social justice movements, in part because they want

to raise awareness about workers' issues and rights, but also in order to build a shared global agenda for social change (Obach 2004a). Examples include the "Teamsters for Turtles" campaign in the "battle of Seattle," which saw unions allied with environmentalists in an effort to create a unified global justice movement (Berg 2003) and the United Students Against Sweatshops, an alliance between college students, NGOs, and labor unions to monitor the labor conditions in factories that produce college apparel (Featherstone 2002).

21. Many of the community groups involved in the coalition—in particular, the main environmental organization—were professionalized, institutionalized organizations, and, as a result, were likely less flexible in their ability and willingness to adopt new frames or deploy more radical, disruptive tactics. Professionalized, bureaucratic groups tend to appeal to more middle class constituents and values (see Croteau 1995). Social movement scholars have also demonstrated that while professionalization increases the capacity and long-term survival of social movement organizations, it can also have a negative effect, hampering an organization's ability to mobilize mass support because of the need to focus on fundraising and management (Jenkins and Eckert 1986; Piven and Cloward 1997; Skocpol 2003).

22. Roth 2003; Milkman 2006.

23. The BCPSEU's ability to gain support from the wider community and build a strong movement against water privatization was strengthened by the fact that the union leaders involved were hired as professional "social movement" staff, who had a specific mandate to work with other social movement organizations on issues of broad appeal. The presence of professional social movement staff also explains why the union would become involved in a campaign in which the outcome would not directly affect their workers.

24. Croteau (1995) argues that while left-wing politics had historically been the domain of the working class, in recent decades liberal, left-of-center social movements have been marked by the absence of white working class individuals. He contends that this is a result of the dominance of "middle class" social movements in leftist politics that no longer consider the working class as a critical source of support.

25. In the mid-1990s the OECD began negotiations for a new global investment treaty, the Multilateral Agreement on Investment (MAI), that would expand international trade and investment regulations to new economic sectors not previously covered by the existing agreements negotiated under the World Trade Organization (WTO), including services such as health and education, trade in currency, and other financial investments such as stocks and bonds, as well as ownership of natural resources (Singer and Orbuch 1997). The MAI negotiations sparked a global opposition movement, in which activists and organizations joined forces to speak out against the potential curtailing of the powers of local and national governments and the threats to environmental protection, and labor and human rights under the agreement. Because of intense public pressure, the MAI talks eventually broke down in 1998 (Public Citizen 1998).

26. Gamson 1990; Amenta, Carruthers, and Zylan 1992; Cress and Snow 2000; Broadbent 2003.

27. Carroll and Ratner 1996; Croteau and Hicks 2003; Van Dyke 2003.

28. The environmental groups in Vancouver were small, locally focused grass-roots organizations that were less professionalized and institutionalized than their counterparts in Stockton. The less formal nature of these organizations could also explain why they were more flexible in shifting frames, adopting new tactics, and working closely with labor organizations.

29. Rose 2000; Obach 2004a; Van Dyke 2003; Mische 2008.

30. In Stockton, the Citizens Environmental Alliance, a national organization that focuses on the link between corporate power and environmental protection, attempted to unite diverse community organizations—unions and environmental organizations in particular—around a common theme of corporate control. Unlike Citizens Action League in Vancouver, however, the Citizens Environmental Alliance was unable to assume a brokerage role because the group lacked ties to local community organizations and thus had not developed the trust and reciprocity needed to assume a bridge-building role.

31. Differences in union culture between the Stockton and Vancouver cases may also be explained by broader organizational cultural differences between private and public sector unions. One key difference between the cases is the nature of the unions involved, with the Utility Workers Local 5 in Stockton largely representing workers in private sector construction and trades, and the BCPSEU in Vancouver mainly representing public sector service employees. The unions that have been most successful incorporating social movement unionism into their organizing practices have largely been service sector unions, representing employees in "post-industrial" jobs (Lopez 2004). The BCPSEU's willingness to focus on broader social issues that resonate beyond the shop floor is likely related to the fact that they are a public sector union and thus have a strong interest in protecting public sector resources from being privatized. At the same time, public sector issues—including health care and water—are more likely to appeal to the broader community because of their importance for the well-being of society, or in the case of water, because it is essential for life.

Research examining differences between private and public sector unions suggests that public sector unions have a greater capacity to mobilize political power and are more likely to engage in social movement unionism than those representing private sector workers. For example, Johnston (1994) argues that because they have diverse goals, public sector unions adopt different cultural forms and tactics than private sector counterparts. He argues that public sector unions assume a political-bureaucratic form (versus the market-oriented form of private sector unions) because of their need to mobilize political organizational resources. Social movement unionism facilitates this process because it enables public sector unions to reach out more broadly to movements and organizations from outside of the labor movement in order to increase their political leverage. At the same time, examples such as the strong social movement emphasis of certain private sector unions—the Service Employees International Union (SEIU) in particular—challenge this argument and suggest that private sector unions benefit both in strength and political influence by shifting to a social movement unionism model (see Milkman 2006).

32. Carroll and Ratner 1996.

33. The variation in union culture in Stockton as compared to Vancouver may reflect differences in union strength between Canada and the United States. Based on statistics compiled in the last decade, the rate of unionization in the United States, nationally, was 12.1 percent of the workforce (Zipperer and Schmitt 2008) as compared with just over 35 per cent in Canada (Lipsit and Meltz 2004). Previous research on labor unions in Canada and the United States demonstrates that Canada's labor unions are relatively strong in terms of political influence as compared to the United States, where the union movement has been in steady decline, with decreasing influence on political decision making and civil society (Riddell 1993; Zuberi 2006). Although there are clear differences between Canada and the United States in terms of union density, these differences should not necessarily be attributed to a divergence in values of the populations of the two countries. While some research has pointed to cross-national variation in values (with Americans being more individualistic and antistatist than Canadians) for explaining differences in union density between Canada and the United States (Lipset 1995), other more recent research suggests that there is little variation between the two countries in terms of support for unionization (Taras and Ponak 2001; Lipset and Meltz 2004). Other research points to differences in labor laws and union organizing rules for explaining differences between the two countries (Chaison and Rose 1985; Taras and Ponak 2001; Johnson 2004).

34. McAdam, Tarrow, and Tilly 2001; Baldassarri and Diani 2007.

35. Baldassarri and Diani 2007; Mische 2008.

36. Some scholars have criticized research on labor revitalization for its lack of conceptual clarity about what constitutes social movement unionism, and the tendency to define all forms of labor revitalization as social movement unionism. These scholars argue that there are important differences between unions engaged in labor revitalization and social movement mobilization both cross-sectorially and cross-nationally, depending on their political orientation and their desire for social change (Park 2007; Shiavone 2007). This includes what Park (2007) calls the distinction between unions that act as "a 'revolutionary agent' against the capitalist system," versus those that act as "a 'social partner' in their pursuit of a collective agreement within a capitalist framework of industrial relations" (Park 2007, 312). The recognition of different forms of labor mobilizations suggests the need to examine more closely and refine the concept of social movement unionism to take into account differences between labor movements that seek broader social change and challenge the structural power of the political economy and those that focus on strengthening union democracy and building coalitions with the wider community in order to improve the conditions of workers.

37. Burawoy 2010, 302. Michael Burawoy argues that what many scholars call new counter-hegemonic global labor movements (see Webster, Lambert, and Bezuindhout 2008 and Evans 2008) are unlikely to lead to the "third great transformation," because they lack a clear and viable vision for an alternative to capitalist hegemony, and thus fail to redirect the system.

38. Lopez 2004.

39. Gottlieb 2002.
40. Obach 2004a.
41. Obach 2004a, 7.

Chapter 7

1. Shiva 2002; Bakker 2010.
2. Oliver and Johnston 2005.
3. Tarrow and McAdam 2005; della Porta and Tarrow 2005.
4. Tarrow (2005) argues that transnational movements not linked to domestic institutions and networks are unlikely to result in sustained mobilization. He points to the importance of "rooted cosmopolitans," social movement actors who mobilize for global causes, but remain connected to networks and resources in their own communities as being the face of the new transnational activism.
5. Pellow 2007.
6. Rose 2000; Lopez 2004; Pellow 2007; Evans 2008.
7. Rose 2000; Evans 2008.
8. Evans 2000; della Porta et al. 2006; Kay 2011.
9. Tarrow 2005.
10. Tarrow and McAdam 2005.
11. Castells 2007.
12. Evans 2008.

References

Adams, Michael. 2003. *Fire and Ice: The United States, Canada and the Myth of Converging Values*. Toronto: Penguin Canada.

Alesina, Alberto, and Edward Glaeser. 2004. *Fighting Poverty in the US and Europe: A World of Difference*. Oxford: Oxford University Press.

Ali, S. Harris. 2004. "A Socio-Ecological Autopsy of the *E. coli* O157:H7 Outbreak in Walkerton, Ontario, Canada." *Social Science & Medicine* 58 (12): 2601–2612.

Amenta, Edwin, Bruce G. Carruthers, and Yvonne Zylan. 1992. "A Hero for the Aged? The Townsend Movement, the Political Mediation Model, and U.S. Old-Age Policy, 1934–1950." *American Journal of Sociology* 98:308–339.

Amenta, Amenta, and Neal Caren. 2004. "The Legislative, Organizational and Beneficiary Consequences of State-Oriented Challengers." In *The Blackwell Companion to Social Movements*, edited by David A. Snow, Sarah A. Soule, and Hanspeter Kriesi, 461–488. Oxford: Blackwell Publishing.

Ancelovici, Marcos. 2002. "Organizing against Globalization: The Case of ATTAC in France." *Politics & Society* 30 (3): 427–463.

Andrews, Kenneth T. 1997. "The Impacts of Social Movements on the Political Process: A Study of the Civil Rights Movement and Black Electoral Politics in Mississippi." *American Sociological Review* 62:800–819.

Andrews, Kenneth T. 2004. *Freedom Is a Constant Struggle: The Mississippi Civil Rights Movement and Its Legacy*. Chicago: Chicago University Press.

Andrews, Kenneth T., and Bob Edwards. 2005. "The Structure of Local Environmentalism." *Mobilization* 10:213–234.

Armstrong, Elizabeth, and Mary Bernstein. 2008. "Culture, Power, and Institutions: A Multi-Institutional Politics Approach to Social Movements." *Sociological Theory* 26:74–99.

Arnold, Thomas Clay. 2008. "The San Luis Valley and the Moral Economy of Water." In *Water Place and Equity*, edited by John M. Whiteley, Helen Ingram, and Richard Warren Perry, 37–68. Cambridge, MA: MIT Press.

Ayres, Jeffrey. M. 2002. "Transnational Political Process and Contention against the Global Economy." In *Globalization and Resistance: Transnational Dimensions of Social Movements*, edited by Jackie G. Smith and Hank Johnston, 191–206. Lanham, MD: Rowman & Littlefield.

Baer, Madeline. 2008. "The Global Water Crisis, Privatization, and the Bolivian Water War." In *Water, Place and Equity*, edited by John M. Whiteley, Helen Ingram, Richard Warren, and Perry Water, 195–224. Cambridge, MA: MIT Press.

Bakker, Karen. 2004. *An Uncooperative Commodity: Privatizing Water in England and Wales*. Oxford: Oxford University Press.

Bakker, Karen. 2005. "Neoliberalizing Nature? Market Environmentalism in Water Supply in England and Wales." *Annals of the Association of American Geographers*. 95 (3): 542–565.

Bakker, Karen. 2007. "The 'Commons' Versus the 'Commodity': Alter-globalization, Anti-privatization and the Human Right to Water in the Global South." *Antipode* 39 (3): 430–455.

Bakker, Karen. 2010. *Privatizing Water: Governance Failure and the World's Urban Water Crisis*. Ithaca, NY: Cornell University Press.

Baldassarri, Delia, and Mario Diani. 2007. "The Integrative Power of Civic Networks." *American Journal of Sociology* 113 (3): 735–780.

Bandy, Joe, and Jackie Smith, eds. 2004. *Coalitions across Borders. Transnational Protest and the Neoliberal Order*. Lanham, MD: Rowman and Littlefield.

Barclay, Scott, Mary Bernstein, and Anna-Maria Marshall, eds. 2009. *Queer Mobilizations: LGBT Activists Confront the Law*. New York: NYU Press.

Barlow, Maude, and Tony Clarke. 2002. *Blue Gold: The Battle Against Corporate Theft of the World's Water*. Toronto: Stoddard Publishing.

Bauer, Carl. 2004. *Siren Song: Chilean Water Law as a Model for International Reform*. Washington, DC: Resources for Future.

Beck, Ulrich. 1992. *Risk Society: Towards a New Modernity*. London: Sage Publications.

Beck, Ulrich. 1996. *The Reinvention of Politics. Rethinking Modernity in the Global Social Order*. Cambridge: Polity Press.

Beck, Ulrich. 1999. *World Risk Society*. Cambridge: Polity Press.

Benford, Robert D. 1993. "You Could Be the Hundredth Monkey": Collective Action Frames and Vocabularies of Motive within the Nuclear Disarmament Movement. *Sociological Quarterly* 34 (2):195–216.

Benford, Robert D. 1997. "An Insider's Critique of the Social Movement Framing Perspective." *Sociological Inquiry* 67:409–430.

Benford, Robert D., and David Snow. 2000. "Framing Processes and Social Movements: An Overview and Assessment." *Annual Review of Sociology* 26:611–639.

Berg, John C., ed. 2003. *Teamsters and Turtles? U.S. Progressive Political Movements in the 21st Century*. Lanham, MD: Rowman & Littlefield.

Berger, Suzanne. 2000. "Globalization and Politics." *Annual Review of Political Science* 3:43–62.

Bernstein, Mary, Anna-Maria Marshall, and Scott Barclay. 2009. "The Challenge of Law: Sexual Orientation, Gender Identity, and Social Movements." In *Queer Mobilizations: LGBT Activists Confront the Law*, edited by Scott Barclay, Mary Bernstein, and Anna-Maria Marshall, 1–17. New York: NYU Press.

Birchfield, Vicki. 2005. "José Bové and the Globalisation Countermovement in France and Beyond: A Polanyian Interpretation." *Review of International Studies* 31 (3): 581–598.

Blatter, Joachim, Helen Ingram, and Pamela Doughman. 2001. "Emerging Approaches to Comprehend Changing Global Contexts." In *Reflections on Water: New Approaches to Transboundary Conflicts and Cooperation*, edited by Helen Ingram and Joachim Blatter, 3–29. Cambridge, MA: MIT Press.

Bluemel, Erik. 2004. "The Implications of Formulating a Human Right to Water." *Ecology Law Quarterly* 31:957.

Bob, Clifford. 2002. "Political Process Theory and Transnational Movements: Dialectics of Protest among Nigeria's Ogoni Minority." *Social Problems* 49 (3): 395–415.

Boelens, Rutgerd, David Getches, and Armando Guevara-Gil. 2010. *Out of the Mainstream: Water Rights, Politics and Identity*. London: Earthscan.

Bolt, Clarence. 1999. *Does Canada Matter? Liberalism and the Illusion of Sovereignty*. Vancouver, BC: Ronsdale Press.

Bonin, Marie-Hélène. 2007. "Stronger Alliances of Unions with Social Movements." *Alternatives*, February 2. http://www.forumdesalternatives.org/en/stronger -alliance-of-unions-with-social-movements.

Bosco, Jennifer. 1994. "Undocumented Immigrants, Economic Justice and Welfare Reform in California." *Georgetown Immigration Law Journal* 8:71.

Bridge, Gavin, and Phil McManus. 2000. "Sticks and Stones: Environmental Narratives and Discursive Regulation in the Forestry and Mining Sectors." *Antipode* 32 (1): 10–47.

British Columbia Public Service Employees Union (BCPSEU). 2001. "Public Control of Water: Too Important to Abandon." Presentation to the Greater Vancouver Regional District Water Committee, April 17.

Broadbent, Jeffrey. 2003. "Movement in Context: Thick Networks and Japanese Environmental Protest." In *Social Movement and Social Networks: Relational Approaches to Collective Action*, edited by Mario Diani and Doug McAdam, 204–232. Oxford: Oxford University Press.

Brulle, Robert J., and J. Craig Jenkins. 2008. "Fixing the Bungled U.S. Environmental Movement." *Contexts* 7 (2): 14–18.

Bulkeley, H., and M. M. Betsill. 2003. *Cities and Climate Change: Urban Sustainability and Global Environmental Governance*. New York: Routledge.

Burawoy, Michael. 2010. "From Polanyi to Pollyanna: The False Optimism of Global Labor Studies." *Global Labour Journal* 1 (2): 301–313.

Burawoy, Michael, Joseph A. Blum, Sheba George, Zsuzsa Gille, Theresa Gowan, Lynne Haney, Maren Klawiter, Steven H. Lopez, Sean O. Riain, and Millie Thayer. 2000. *Global Ethnography: Forces, Connections, and Imaginations in a Postmodern World*. Berkeley: University of California Press.

Burke, Garance. 2004. "Suit Tests the Waters on Control of Utilities: The Stockton Case Could Set a Precedent for the Privatization of Municipal Systems." *Sacramento Bee*, July 4.

Burstein, Paul. 1999. "Social Movements and Public Policy." In *How Social Movements Matter*, edited by Marco Guigni, Doug McAdam, and Charles Tilly, 3–21. Minneapolis: University of Minnesota Press.

Burnstein, Paul, Rachel Einwohner, and Jocelyn Hollander. 1995. "The Success of Social Movements: A Bargaining Perspective." In *The Politics of Social Protest: Comparative Perspectives on States and Social Movements*, edited by J. Craig Jenkins and Bert Klandermans, 275–295. Minneapolis: University of Minnesota Press.

Capek, S. M. 1993. "The 'Environmental Justice' Frame: A Conceptual Discussion and Application." *Social Problems* 40:5–24.

Carroll, William. 1993. "Canada in the Crisis: Transformations in Capital Structure and Political Strategy." In *Restructuring Hegemony in the Global Political Economy: The Rise of Transnational Neo-liberalism in the 1980s*, edited by Henk Overbeek, 216–245. London: Routledge.

Carroll, William K., and R. S. Ratner. 1996. "Master Framing and Cross-Movement Networking in Contemporary Social Movements." *Sociological Quarterly* 37 (4): 601–625.

Castells, Manuel. 2003. "Global Informational Capitalism." In *The Global Transformations Reader: An Introduction to the Globalization Debate,* edited by David Held and Anthony McGrew, 311–334. Cambridge: Polity.

Castells, Manuel. 2007. "Communication, Power and Counter-power in the Network Society." *International Journal of Communication* 1:238–266.

Castro, José Estaban. 2007. "Poverty and Citizenship: Sociological Perspectives on Water Services and Public-Private Participation." *Geoforum* 38:753–771.

Catley-Carlson, Margaret. 2003. "Working for Water." In *Whose Water Is It? The Unquenchable Thirst of a Water-Hungry World*, edited by Bernadette McDonald and Douglas Jehl, 65–76. Washington, DC: National Geographic Society.

Catton, William, and Riley Dunlap. 1978. "Environmental Sociology: A New Paradigm." *American Sociologist* 13:41–49.

Chaison, Gary N., and Joseph B. Rose. 1985. "The State of the Unions: United States and Canada." *Journal of Labor Research* 6 (1): 97–111.

Chun, Jennifer Jihye. 2009. *Organizing at the Margins: The Symbolic Politics of Labor in South Korea and the United States*. Ithaca, NY: Cornell University Press.

Citizens Action League. 2001. Fact Sheet: Water and Trade. March.

Clarke, Tony. 2003. "Water Privateers." *Alternatives Journal* 29 (2): 10–15.

Clawson, Dan. 2003. *The Next Upsurge: Labor and the New Social Movements*. Ithaca, NY: Cornell University Press.

Conca, Ken. 2005. *Governing Water: Contentious Transnational Politics and Global Institution Building*. Cambridge, MA: MIT Press.

Conway, Janet M. 2004. *Identity, Place, and Knowledge: Social Movements Contesting Globalization*. Halifax: Fernwood Books.

Cress, Daniel M., and David A. Snow. 2000. "The Outcomes of Homeless Mobilization: The Influence of Organization, Disruption, Political Mediation, and Framing." *American Journal of Sociology* 105 (4): 1063–1104.

Croteau, David. 1995. *Politics and the Class Divide. Working People and the Middle Class Left*. Philadelphia: Temple University Press.

Croteau, David, and Lyndsi Hicks. 2003. "Coalition Framing and the Challenge of a Consonant Frame Pyramid: The Case of a Collaborative Response to Homelessness." *Social Problems* 50 (2): 251–272.

Dauvergne, Peter. 2008. *The Shadows of Consumption: Consequences for the Global Environment*. Cambridge, MA: MIT Press.

della Porta, Donatella, Massimiliano Andretta, Lorenzo Mosca, and Herbert Reiter. 2006. *Globalization from Below: Transnational Activists and Protest Networks*. Minneapolis: University of Minnesota Press.

della Porta, Donatella, and Sidney Tarrow, eds. 2005. *Transnational Protest and Global Activism*. New York: Rowman and Littlefield.

Diani, Mario. 1995. *Green Networks: A Structural Analysis of the Italian Environmental Movement*. Edinburgh: Edinburgh University Press.

Diani, Mario. 1996. "Linking Mobilization Frames and Political Opportunities: Insights from Regional Populism in Italy." *American Sociological Review* 61 (6): 1053–1069.

Diani, Mario. 2003. "Networks and Social Movements: A Research Programme." In *Social Movement and Social Networks: Relational Approaches to Collective Action*, edited by Mario Diani and Doug McAdam, 299–319. Oxford: Oxford University Press.

Diani, Mario, and Ivano Bison. 2004. "Organizations, Coalitions, and Movements." *Theory and Society* 33:281–309.

Diani, Mario, and Doug McAdam, eds. 2003. *Social Movements and Networks: Relational Approaches to Collective Action*. Oxford: Oxford University Press.

Diani, Mario, and Elisa Rambaldo. 2007. "Still the Time of Environmental Movements? A Local Perspective." *Environmental Politics* 16 (5): 765–784.

Dowie, Mark. 1996. *Losing Ground: Environmentalism at the Close of the Twentieth Century*. Cambridge, MA: MIT Press.

Dryzek, John. S. 2006. *Deliberative Global Politics: Discourse and Democracy in a Divided World*. Cambridge: Polity.

Dryzek, John S., David Downes, Christian Hunold, David Schlosberg, and Hans-Kristian Hernes. 2003. *Green States and Social Movements: Environmentalism in the United States, United Kingdom, Germany, and Norway*. Oxford: Oxford University Press.

Dunlap, Riley, and William Catton. 1994. "Struggling with Human Exceptionalism." *American Sociologist* 25 (1): 5–30.

Dunlap, Riley E., and William Michelson, eds. 2002. *Handbook of Environmental Sociology*. Westport, CT: Greenwood Press.

Dupuis, Melanie E., and Daniel Block. 2008. Sustainability and Scale: US Milk Market Orders as Relocalization Policy. *Environment & Planning A* 40 (8): 1987–2005.

DuPuis, E. Melanie, and David Goodman. 2005. "Should We Go 'Home' to Eat? Towards a Reflexive Politics of Localism." *Regional Studies* 21 (3): 359–371.

Earthjustice. 2008. "New Biological Opinion Will Protect San Francisco Bay-Delta: State and Federal Water Projects Operators Must Protect Native Fish from Extinction," December 15. http://www.calsport.org/12-15-08d.htm.

Epp, Charles. R. 1998. *The Rights Revolution: Lawyers, Activists and Supreme Courts in Comparative Perspective*. Chicago: University of Chicago Press.

Espeland, Wendy Nelson. 1998. *The Struggle for Water: Politics, Rationality and Identity in the American Southwest*. Chicago: University of Chicago Press.

Estabrook, Thomas. 2007. *Labor and Environmental Coalitions: Lessons from a Louisiana Petrochemical Region*. Amityville, NY: Baywood Publishing.

Evans, Peter. 1995. *Embedded Autonomy: States and Industrial Transformation*. Princeton, NJ: Princeton University Press.

Evans, Peter. 2000. "Fighting Marginalization with Transnational Networks: Counterhegemonic Globalization." *Contemporary Sociology* 29:230–241.

Evans, Peter. 2008. "Is an Alternative Globalization Possible?" *Politics & Society* 36 (2): 271–305.

Evans, Rhonda, and Tamara Kay. 2008. "How Environmentalists 'Greened" Trade Policy Under NAFTA: Strategic Action and the Architecture of Field Overlap." *American Sociological Review* 73 (6): 970–991.

Eyerman, Ron, and Andrew Jamison. 1991. *Social Movements: A Cognitive Approach*. University Park: Pennsylvania State University Press.

Fantasia, Rick, and Kim Voss. 2004. *Hard Work: Remaking the American Labor Movement*. Berkeley: University of California Press.

Featherstone, Liza. 2002. *Students Against Sweatshops: The Making of a Movement*. London: Verso.

Ferree, Myra Marx, William Anthony Gamson, Jurgen Gerhards, and Dieter Rucht. 2002. *Shaping Abortion Discourse: Democracy and the Public Sphere in Germany and the United States*. Cambridge: Cambridge University Press.

Fisher, Dana, Kevin Stanley, David Berman, and Gina Neff. 2005. "How Do Organizations Matter? Mobilization and Support for Participants at Five Globalization Protests." *Social Problems* 52 (1): 102–121.

Fiss, Peer C., and Paul Hirsch. 2005. "The Discourse of Globalization: Framing and Sensemaking of an Emergent Concept." *American Sociological Review* 70:29–52.

Gamson, William A. 1990. *The Strategy of Social Protest*. 2nd ed. Belmont, CA: Wadsworth Publishing.

Gamson, William A., and David S. Meyer. 1996. "Framing Political Opportunity." In *Comparative Perspectives on Social Movements: Opportunities, Mobilizing Structures and Framing*, edited by Doug McAdam, John D. McCarthy, and Meyer N. Zald, 275–290. New York: Cambridge University Press.

Ganz, Marshall. 2000. "The Paradox of Powerlessness: Strategic Capacity in the Unionization of California Agriculture, 1959–1966." *American Journal of Sociology* 105:1003–1062.

Garcia, Ruben J. 1995. "Critical Race Theory and Proposition 187: The Racial Politics of Immigration Law." *Chicano-Latino Law Review* 17:118.

Gerber, Elizabeth R. 1999. *The Populist Paradox: Interest Group Influence and the Promise of Direct Legislation*. Princeton, NJ: Princeton University Press.

Gerber, Elizabeth R., and Justin H. Phillips. 2005. "Evaluating the Effects of Direct Democracy on Public Policy: California's Urban Growth Boundaries." *American Politics Research* 33 (2): 310–330.

Giddens, Anthony. 1990. *The Consequences of Modernity*. Cambridge: Polity Press.

Giddens, Anthony. 2009. *The Politics of Climate Change*. Cambridge: Polity.

Giugni, Marco G. 1998. "Was It Worth the Effort? The Outcomes and Consequences of Social Movements." *Annual Review of Sociology* 98:371–393.

Giugni, Marco. 2004. *Social Protest and Policy Change: Ecology, Anti-Nuclear and Peace Movements in Comparative Perspective*. Lanham, MD: Rowman & Littlefield.

Giugni, Marco, Doug McAdam, and Charles Tilly, eds. 1999. *How Social Movements Matter*. Minneapolis: University of Minnesota Press.

Gleick, Peter H. 1996. "Basic Water Requirements for Human Activities: Meeting Basic Needs." *Water International* 21 (2): 83–92.

Gleick, Peter. 2002. *The World's Water: A Biennial Report on Freshwater Resources 2002–2003*. Washington, DC: Island Press.

Goffman, Erving. 1974. *Frame Analysis*. New York: Harper.

Goldman, Michael. 2007. "How 'Water for All!' Policy Became Hegemonic: The Power of the World Bank and Its Transnational Policy Networks." *Geoforum* 38:786–800.

Goodwin, Jeff. 2001. *No Other Way Out: States and Revolutionary Movements, 1945–1991*. New York: Cambridge University Press.

Goodwin, Jeff, and James M. Jasper. 1999. "Caught in a Winding, Snarling Vine: The Structural Bias of Political Process Theory." *Sociological Forum* 14 (1): 27–54.

Goodwin, Jeff, and James M. Jasper. 2004. *Rethinking Social Movements: Structure, Meaning, and Emotion*. Lanham, MD: Rowman & Littlefield.

Goodwin, Jeff, James M. Jasper, and Francesca Polletta. 2001. *Passionate Politics: Emotions and Social Movements*. Chicago: University of Chicago Press.

Gordon, Tracy M. 2004. *The Local Initiative in California*. San Francisco: Public Policy Institute of California.

Gottlieb, Robert. 1989. *A Life of Its Own: The Politics and Power of Water*. San Diego: Harcourt Brace Jovanovich.

Gottlieb, Robert. 2002. *Environmentalism Unbound: Exploring New Pathways for Change*. Cambridge, MA: MIT Press.

Gould, Kenneth A., Tammy L. Lewis, and J. Timmons Roberts. 2004. "Blue-Green Coalitions: Constraints and Possibilities in the Post 9-11 Political Environment." *Journal of World-systems Research* 10 (1): 91–116.

Gould, Kenneth A., David N. Pellow, and Allan Schnaiberg. 2008. *The Treadmill of Production: Injustice and Unsustainability in the Global Economy*. Boulder, CO: Paradigm Publishers.

Gould, Kenneth A., Allan Schnaiberg, and Adam S. Weinberg. 1996. *Local Environmental Struggles: Citizen Activism in the Treadmill of Production*. New York: Cambridge University Press.

Greater Vancouver Regional District (GVRD). 2001. "GVWD Decides Against Design-Build-Operate Arrangement for Construction of Drinking Water Filtration Facilities." News release, June 29.

Greater Vancouver Regional District (GVRD). 1999. Minutes of the Administration Board Meeting, May 28.

Greater Vancouver Regional District (GVRD). 2001. In-Camera Meeting Minutes of the Administration Board, June 28.

Greater Vancouver Regional District (GVRD). 2001. Meeting Minutes of the Water Committee, July 13.

Greider, Thomas, and Lorraine Garkovich. 1994. "Landscapes: The Social Construction of Nature and the Environment." *Rural Sociology* 59 (1): 1–24.

Guidry, John A., Michael D. Kennedy, and Mayer N. Zald, eds. 2000. *Globalizations and Social Movements: Culture, Power, and the Transnational Public Sphere*. Ann Arbor: University of Michigan Press.

Hackworth, Jason. 2007. *The Neoliberal City: Governance, Ideology, and Development in American Urbanism*. Ithaca, NY: Cornell University Press.

Hajer, Maarten. 2003. "Policy without Polity? Policy Analysis and the Institutional Void." *Policy Sciences* 36:175–195.

Hamel, Pierre, Henri Lustiger-Thaler, Jan Nederveen Pieterse, and Sasha Roseneil, eds. 2001. *Globalization and Social Movements*. New York: Palgrave Press.

Hannigan, John. 1995. *Environmental Sociology: A Social Constructionist Perspective*. New York: Routledge.

Harvey, David. 2005. *A Brief History of Neoliberalism*. New York: Oxford University Press.

Hecksher, Charles, and David Palmer. 1993. "Associational Movements and Employment Rights: An Emerging Paradigm?" In *Research in the Sociology of Organizations: Special Issue on Labor Relations and Unions 12*, edited by Samuel B. Bacharach, Ronald Seeber, and David Walsh. Greenwich, CT: JAI Press.

Herod, Andrew. 2001. "Labor Internationalism and the Contradictions of Globalization: Or, Why the Local Is Still Important in a Global Economy." *Antipode* 33 (3): 407–426.

Hess, David J. 2009. *Localist Movements in a Global Economy: Sustainability, Justice, and Urban Development in the United States*. Cambridge, MA: MIT Press.

Hilson, Chris. 2002. "New Social Movements: The Role of Legal Opportunity." *Journal of European Public Policy* 9 (2): 238–255.

HoSang, Daniel Martinez. 2010. *Racial Propositions: Ballot Initiatives and the Making of Postwar California*. Berkeley: University of California Press.

Inglehart, Ronald. 2008. "Changing Values among Western Publics from 1970 to 2006." *West European Politics* 31 (1–2): 130–146.

Ingram, Helen, John M. Whiteley, and Richard Perry. 2008. "The Importance of Equity and the Limits of Efficiency in Water Resources." In *Water Place and Equity*, edited by John M. Whiteley, Helen Ingram, and Richard Warren Perry, 1–32. Cambridge, MA: MIT Press.

Jasper, James M. 1997. *The Art of Moral Protest*. Chicago: University of Chicago Press.

Jenkins, Craig. J., and Bert Klandermans, eds. 1995. *The Politics of Social Protest: Comparative Perspectives on States and Social Movements*. Minneapolis: University of Minnesota Press.

Jenkins, J. Craig, and Craig M. Eckert. 1986. "Channeling Black Insurgency: Elite Patronage and Professional Social Movement Organizations in the Development of the Black Movement." *American Sociological Review* 51 (6):812–829.

Jenkins, J. Craig, David Jacobs, and Jon Agnone. 2003. "Political Opportunities and African-American Protest: 1948–1997." *American Journal of Sociology* 109 (2):277–303.

Johnson, Erik, and John D. McCarthy. 2005. "Mobilization of the Global and U.S. Environmental Movements." In *Transnational Protest and Global Activism*, edited by Donatella della Porta and Sidney Tarrow, 71–94. New York: Rowman & Littlefield.

Johnston, Hank. 2005. "Comparative Frame Analysis." In *Frames of Protest: Social Movements and the Framing Perspective*, edited by Hank Johnston and John A. Noakes, 237–260. Lanham, MD: Rowman & Littlefield.

Johnston, Hank, and John Noakes, eds. 2005. *Frames of Protest: Social Movements and the Framing Perspective*. Lanham, MD: Rowman & Littlefield.

Johnston, Paul. 1994. *Success While Others Fail: Social Movement Unionism and the Public Workplace*. Ithaca, NY: Cornell University Press.

Joppke, Christian. 1993. *Mobilizing Against Nuclear Energy: A Comparison of Germany and the United States*. Berkeley: University of California Press.

Josselin, Daphné. 2007. "From Transnational Protest to Domestic Political Opportunities Insights from the Debt Cancellation Campaign." *Social Movement Studies* 6 (1): 21–38.

Juris, Jeffrey S. 2004. "Networked Social Movements: Global Movements for Global Justice." In *The Network Society: a Cross-Cultural Perspective*, edited by Manuel Castells, 341–362. Cheltenham, UK: Edward Elgar.

Kay, Tamara. 2005. "Labor Transnationalism and Global Governance: The Impact of NAFTA on Transnational Labor Relationships in North America." *American Journal of Sociology* 111 (3): 715–756.

Kay, Tamara. 2011. *NAFTA and the Politics of Labor Transnationalism*. New York: Cambridge University Press.

Keck, Margaret E., and Kathryn Sikkink. 1998. *Activists beyond Borders: Advocacy Networks in International Politics*. Ithaca, NY: Cornell University Press.

Keil, Roger. 2002. "New Geographies of Power, Exclusion and Injustice 'Common–Sense' Neoliberalism: Progressive Conservative Urbanism in Toronto, Canada." *Antipode* 34 (3):578–601.

Khagram, Sanjeev. 2004. *Dams and Development: Transnational Struggles for Water and Power*. Ithaca, NY: Cornell University Press.

Khagram, Sanjeev, James V. Riker, and Kathryn Sikkink, eds. 2002. *Restructuring World Politics: Transnational Social Movements, Networks, and Norms*. Minneapolis: University of Minnesota Press.

Klandermans, Bert, and Suzanne Staggenborg, eds. 2002. *Methods of Social Movement Research*. Minneapolis: University of Minnesota Press.

Klein, Greg. 2001. "GVRD Backs Off on Private Water Control." *North Shore News*, July 2.

Kranz, Adam. 2004. "As Election Approaches American Public Shows Overwhelming Support for Clean Water Funding." National Association of Clean Water Agencies, February 26. http://archive.nacwa.org/advocacy/releases/022604.cfm.

Lees, Nancy. 2001. "Residents Rally against GVRD Water Privatization." *Burnaby Now*, June 17.

Leitner, Helga, Eric Sheppard, Kristin Sziarto, and Ananthakrishna Maringanti. 2007. "Contesting Urban Futures: Decentering Neoliberalism." In *Contesting Neoliberalism: Urban Frontiers*, edited by Helga Leitner, Jamie Peck, and Eric Sheppard, 1–25. New York: Guilford Press.

Lipset, Seymour Martin. 1990. *Continental Divide: The Values and Institutions of the United States and Canada*. New York: Routledge.

Lipset, Seymour Martin. 1995. "Trade Union Exceptionalism: The United States and Canada." *Annals of the American Academy of Political and Social Science* 538:115–130.

Lipset, Seymour Martin, and Noah M. Meltz. 2004. *The Paradox of American Unionism*. Ithaca, NY: Cornell University Press.

Loftus, Alex, and Fiona Lumsden. 2008. "Reworking Hegemony in the Urban Landscape." *Transactions of the Institute of British Geographers* 33 (1): 109–126.

Lopez, Steven Henry. 2004. *Reorganizing the Rust Belt: An Inside Study of the American Labor Movement*. Berkeley: University of California Press.

Low, Nicholas, and Brendan Gleeson. 1998. *Justice, Society, and Nature: An Exploration of Political Ecology*. London: Routledge.

Luntz Research Companies. 2005. "New Poll: Americans Overwhelmingly Support Federal Trust Fund to Guarantee Clean and Safe Water." Alexandria, VA, March 3.

Luntz, Frank. 2009. "Infrastructure: It's Job 1 to Americans: A Poll Finds Near Unanimous Support for Rebuilding." *Los Angeles Times*, January 23, http://www.latimes.com/news/opinion/commentary/la-oe-luntz23-2009jan23,0,2761866.story.

Magnusson, Warren. 2005. "Urbanism, Cities and Local Self-government." *Canadian Public Administration* 48 (1): 96–123.

Martin, Isaac William. 2009. "Proposition 13 Fever: How California's Tax Limitation Spread." *California Journal of Politics and Policy* 1 (1): 1–17.

Matsusaka, John G. 2004. *For the Many or the Few: The Initiative, Public Policy, and American Democracy*. Chicago: University of Chicago Press.

McAdam, Doug. 1996. "Conceptual Origins, Current Problems, Future Directions." In *Comparative Perspectives on Social Movements: Opportunities, Mobilizing Structures and Framing*, edited by Doug McAdam, John D. McCarthy, and Meyer N. Zald, 23–40. New York: Cambridge University Press.

McAdam, Doug. 1999. *Political Process and the Development of Black Insurgency, 1930–1970*. Chicago: University of Chicago Press.

McAdam, Doug. 2003. "Beyond Structural Analysis: Toward a More Dynamic Understanding of Social Movements." In *Social Movements and Networks: Relational Approaches to Collective Action*, edited by Mario Diani and Doug McAdam, 281–298. Oxford: Oxford University Press.

McAdam, Doug, John D. McCarthy, and Meyer N. Zald, eds. 1996. *Comparative Perspectives on Social Movements: Opportunities, Mobilizing Structures and Framing*. New York: Cambridge University Press.

McAdam, Doug, Sidney Tarrow, and Charles Tilly. 2001. *Dynamics of Contention*. New York: Cambridge University Press.

McAdam, Doug, Charles Tilly, and Sidney Tarrow. 2008. "Progressive Polemics: Reflections on Four Stimulating Commentaries." *Qualitative Sociology* 31:361–367.

McCaan, Michael. 2004. "Law and Social Movements." In *The Blackwell Companion to Law and Society*, edited by Austin Sarat, 506–522. Malden, MA: Blackwell Publishing.

McCarthy, John D., and Mayer N. Zald. 1977. "Resource Mobilization and Social Movements: A Partial Theory." *American Journal of Sociology* 82:1212–1241.

McNaghten, Phil, and John Urry. 1998. *Contested Natures*. Thousand Oaks, CA: Sage Publications.

McNaghten, P., R. Grove-White, M. Jacobs, and B. Wynne. 1995. *Public Perceptions and Sustainability in Lancashire: Indicators, Institutions, Participation*. Lancaster, UK: Centre for the Study of Environmental Change.

Meinzen-Dick, Ruth S., and Claudia Ringler. 2008. "Water Reallocation: Drivers, Challenges, Threats, and Solutions for the Poor." *Journal of Human Development* 9 (1): 47–64.

Melucci, Alberto. 1988. "Getting Involved: Identity and Mobilization in Social Movements." *International Social Movement Research* 1:329–348.

Melucci, Alberto. 1996. *Challenging Codes: Collective Action in the Information Age*. Cambridge: Cambridge University Press.

Meyer, David S. 2003. "Political Opportunity and Nested Institutions." *Social Movement Studies* 2 (1): 17–35.

Meyer, David S. 2004. "Protest and Political Opportunities." *Annual Review of Sociology* 30:125–145.

Meyer, David S., and Suzanne Staggenborg. 1996. "Movements, Countermovements, and the Structure of Political Opportunity." *American Journal of Sociology* 101:1628–1660.

Milkman, Ruth. 2006. *L.A. Story: Immigrant Workers and the Future of the U.S. Labor Movement*. New York: Russell Sage Foundation.

Miller, Jodi. 1999. "Democracy in Free Fall: The Use of Ballot Initiatives to Dismantle State-Sponsored Affirmative Action Programs." *Annual Survey of American Law*: 1–23.

Mische, Ann. 2008. *Partisan Publics: Communication and Contention across Brazilian Youth Activist Networks*. Princeton, NJ: Princeton University Press.

Morgan, Bronwen. 2011. *Water on Tap: Rights and Reglutions in the Transnational Governance of Urban Water Systems*. Cambridge: Cambridge University Press.

Moulton, Lynne, and Helmut K. Anheim. 2000. "Public-Private Partnerships in the United States: Historical Patterns and Current Trends." In *Public-Private Partnerships: Theory and Practice in International Perspective*, edited by Stephen P. Osborne, 105–119. London: Routledge.

Mullin, Meagan. 2009. *Governing the Tap: Special District Governance and the New Local Politics of Water*. Cambridge, MA: MIT Press.

Munck, Ronaldo. 2002. *Globalisation and Labour: The New "Great Transformation."* London: Zed Books.

Nanos Research. 2008. "Nanos Poll: Views on Canadian Municipal Public Services." Ottawa, May 11.

Nissen, Bruce. 2004. "The Effectiveness and Limits of Labor- Community Coalitions: Evidence from South Florida." *Labor Studies Journal* 29 (1): 67–89.

Obach, Brian. 2004a. *Labor and the Environmental Movement: The Quest for Common Ground*. Cambridge: MIT Press.

Obach, Brian. 2004b. "New Labor: Slowing the Treadmill of Production?" *Organization & Environment* 17 (3): 337–354.

O'Brian, Richard. 1992. *Global Financial Integration: The End of Geography*. Washington, DC: Council on Foreign Relations Press.

Olesen, Thomas. 2005. "The Uses and Misuses of Globalization in the Study of Social Movements." *Social Movement Studies* 4 (1): 49–63.

Oliver, Pamela E., and Hank Johnston. 2005. "What a Good Idea! Ideologies and Frames in Social Movement Research." In *Frames of Protest: Social Movements and the Framing Perspective*, edited by Hank Johnston and John A. Noakes, 185–203. Lanham, MD: Rowman & Littlefield.

Olivera, Oscar, and Tom Lewis. 2004. *Cochabamba! Water Rebellion in Bolivia*. Cambridge, MA: South End Press.

Park, Mi. 2007. "South Korean Trade Union Movement at the Crossroads: A Critique of 'Social-Movement' Unionism." *Critical Sociology* 33:311–344.

Peck, Jamie, and Adam Tickell. 2002. "Neoliberalizing Space." *Antipode* 34:380–404.

Pellow, David. 2001. "Environmental Justice and the Political Process: Movements, Corporations, and the State." *Sociological Quarterly* 42 (1): 47–67.

Pellow, David Naguib. 2007. *Resisting Global Toxics: Transnational Movements for Environmental Justice.* Cambridge: MIT Press.

Pellow, David N., and Lisa Sun-Hee Park. 2002. *The Silicon Valley of Dreams: Environmental Injustice, Immigrant Workers, and the High-Tech Global Economy.* New York: New York University Press.

Perkel, Colin N. 2002. *Well of Lies: The Walkerton Water Tragedy.* Toronto: McLelland & Stewart.

Perry, Richard, Helen Ingram, and John Whiteley, eds. 2008. *Water, Place and Equity.* Cambridge, MA: MIT Press.

Piven, Frances Fox, and Richard A. Cloward. 1977. *Poor People's Movements: Why They Succeed, How They Fail.* New York: Vintage Books.

Piven, Frances Fox, and Richard A. Cloward. 1997. *The Breaking of the American Social Compact.* New York: The New Press.

Presas, Luciana, and Arthur P. J. Mol. 2006. "Greening Transnational Buildings: Between Global Flows and Local Places." In *Governing Environmental Flows: Global Challenges to Social Theory,* edited by Gert Spaargaren, Arthur P. J. Mol, and Frederick H. Buttel, 303–326. Cambridge, MA: MIT Press.

Prudham, Scott. 2003. "Poisoning the Well: Neoliberalism and the Contamination of Municipal Water in Walkerton, Ontario." *Geoforum* 35:343–359.

Public Citizen. "Everything You Wanted to Know about the MAI—But Didn't Know to Ask . . ." http://www.citizen.org/trade/issues/mai/articles.cfm?ID=5626.

Public Citizen. 2002. "Water Privatization in Stockton, California: Backgrounder, November 4. http://www.citizen.org/documents/UpdatedBackgrounder_20021104.pdf.

Public Citizen. 2003. "Stockton Overrides Public Right to Vote on Controversial Water Deal." News release, February 19.

"RealtyTrac: Stockton Had Nation's Highest Q2 Foreclosure Rate; Oakland Was No. 8." 2008. *San Francisco Business Times,* July 25. http://www.bizjournals.com/eastbay/stories/2008/07/21/daily76.html.

Reese, Ellen, and Garnett Newcombe. 2003. "Income Rights, Mothers' Rights, or Workers' Rights? Collective Action Frames, Organizational Ideologies, and the American Welfare Rights Movement." *Social Problems* 50 (2): 294–318.

Riddel, W. Craig. 1993. "Unionization in Canada and the United States: A Tale of Two Countries." In *Small Differences that Matter: Labor Markets and Income Maintenance in Canada and the United States,* edited by David Card and Richard B. Freeman, 109–149. Chicago: University of Chicago Press.

Roberts, J. Timmons, and Melissa Toffolon-Weiss. 2001. *Chronicles from the Environmental Justice Frontline.* New York: Cambridge University Press.

Rose, Fred. 2000. *Coalitions Across the Class Divide: Lessons from the Labor, Peace, and Environmental Movements.* Ithaca, NY: Cornell University Press.

Roth, Silke. 2003. *Building Movement Bridges: The Coalition of Labor Union Women.* Westport, Connecticut: Praeger.

Rothfeder, Jeffrey. 2001. *Every Drop for Sale: Our Desperate Battle Over Water*. New York: Jeremy P. Tarcher.

Samuel, Terrance. 2009. "A Good Working Environment." *The American Prospect*, February 26. http://www.prospect.org/cs/articles?article=a_good_working _environment.

Sassen, Saskia. 2000. "The Global City: Strategic Site/New Frontier." In *Democracy, Citizenship and the Global City*, edited by Engin F. Isin, 48–61. London: Routledge.

Sassen, Saskia. 2008. "Neither Global nor National: Novel Assemblages of Territory, Authority and Rights." *Ethics and Global Politics* 1 (1–2):61–79.

Satterfield, Terre. 2002. *Anatomy of a Conflict: Identity, Knowledge, and Emotion in Old-Growth Forests*. Vancouver, BC: UBC Press.

Schnaiberg, Allan, and Kenneth A. Gould. 2000. *Environment and Society: The Enduring Conflict*. West Caldwell, NJ: Blackburn Press.

Schoenberger, Erica. 2003. "The Globalization of Environmental Management: International Investment in the Water, Wastewater and Solid Waste Industries." In *Remaking the Global Economy*, edited by Jamie Peck and Henry Wai-chung Yeung, 83–100. London and Thousand Oaks, CA: Sage

Schouten, Marco, and Klaus Schwarz. 2006. "Water as a Political Good: Implications for Investments." *International Environmental Agreement: Politics, Law and Economics* 6:407–421.

Schurman, Rachel, and William Munro. 2006. "Ideas, Thinkers and Social Networks: The Process of Grievance Construction in the Anti-Genetic Engineering Movement." *Theory and Society*. 35:1–38.

Schurman, Rachel, and William Munro. 2009. "Targeting Capital: A Cultural Economy Approach to Understanding the Efficacy of Two Anti-Genetic Engineering Movements." *American Journal of Sociology* 115 (1): 155–202.

Shiavone, Michael. 2007. "Moody's Account of Social Movement Unionism: An Analysis." *Critical Sociology* 33 (1): 279–309.

Shiva, Vandana. 2002. *Water Wars: Privatization, Pollution, and Profit*. Cambridge, MA: South End Press.

Shrybman, Steven. 2002. *Thirst for Control: New Rules in the Global Water Grab. The Blue Planet Project*. Ottawa: The Council of Canadians.

Siders, David. 2007. "$600 M Water Deal Runs Dry." *Stockton Record*, July 18.

Simpson, Scott. 2001 "GVRD to Make Water Privatization Plans Public." *Vancouver Sun*, April 28.

Singer, Thomas, Paul Orbuch, and Robert Stumberg. 1997. "Multilateral Agreement on Investment: Potential Effects on State & Local Government." Western Governors' Association.

Skocpol, Theda. 2003. *Diminished Democracy. From Membership to Management in American Civic Life*. Norman: University of Oklahoma Press.

Skoloff, Brian. 2003. "Stockton Water Deal Stirs Privatization Ire." *Associated Press*, March 30.

Smith, Jackie. 2002. "Globalizing Resistance: The Battle of Seattle and the Future of Social Movements." In *Globalization and Resistance: Transnational Dimensions of Social Movements*, edited by J. Smith and H. Johnston, 207–228. Lanham, MD: Rowman & Littlefield.

Smith, Jackie, Charles Chatfield, and Ron Pagnucco. 1997. *Transnational Social Movements and Global Politics. Solidarity Beyond the State.* Syracuse, NY: Syracuse University Press.

Smith, Jackie, and Hank Johnston, eds. 2002. *Globalization and Resistance: Transnational Dimensions of Social Movements.* Lanham, MD: Rowman & Littlefield.

Snitow, Alan, and Deborah Kaufman. 2007. *Thirst: Fighting the Corporate Theft of Water.* San Francisco: Jossey-Bass.

Snow, David A., and Robert D. Benford. 1988. "Ideology, Frame Resonance, and Participant Mobilization." *International Social Movement Research* 1:197–217.

Snow, David A., E. Burke Rochford, Jr., Steven K. Worden, and Robert D. Benford. 1986. "Frame Alignment Processes, Micromobilization and Movement Participation." *American Sociological Review* 51:464–481.

Snow, David A., and Danny Trom. 2002. "The Case Study and the Study of Social Movements." In *Methods of Social Movement Research*, edited by Bert Klandermans and Suzanne Staggenborg, 146–172. Minneapolis: University of Minnesota Press.

Society Promoting Environmental Conservation. 1999. "SPEC to Cooperate with GVRD on No-logging Policy for Vancouver Water Supply." News release, December 13.

Staggenborg, Suzanne. 1986. "Coalition Work in the Pro-Choice Movement: Organizational and Environmental Opportunities and Obstacles." *Social Problems* 33 (5): 374–390.

Staggenborg, Suzanne. 1988. "The Consequences of Professionalization and Formalization in the Pro-Choice Movement." *American Sociological Review* 53 (4): 585–605.

Staggenborg, Suzanne. 1991. *The Pro-Choice Movement: Organization and Activism in the Abortion Conflict.* New York: Oxford University Press.

Staggenborg, Suzanne, and Josée Lecomte. 2009. "Social Movement Campaigns: Mobilization and Outcomes in the Montreal Women's Movement." *Mobilization* 14 (2): 405–422.

Standing, Guy. 2002. *Beyond the New Paternalism: Basic Security as Equality.* London: Verso.

Stiglitz, Joseph E. 2002. *Globalization and its Discontents.* New York: W.W. Norton & Company.

Stillerman, Joel. 2003. "Transnational Activist Networks and the Emergence of Labor Internationalism in the NAFTA Countries." *Social Science History* 27 (4): 577–601.

Stockton Record. 2007. "Stockton Water Privatization Failure Not a Total Waste." July 23.

Swidler, Ann. 1986. "Culture in Action: Symbols and Strategies." *American Sociological Review* 51: 273–286.

Swyngedouw, Eric. 2004. *Social Power and the Urbanization of Water: Flows of Power*. Oxford: Oxford University Press.

Tady, Megan. "A Win in the Water War." 2007. *In These Times*, August 1.

Tal, Benjamin. 2006. "Tapping into Water." *CIBC World Markets Occasional Report #59*.

Taras, Daphne Gottlieb, and Allen Ponak. 2001. "Mandatory Agency Shop Laws as an Explanation of Canada-U.S. Union Density Divergence." *Journal of Labor Research* 22 (3):541–568.

Tarrow, Sidney. 1998. *Power in Movement: Social Movements and Contentious Politics*. New York: Cambridge University Press.

Tarrow, Sidney. 2002. "From Lumping to Splitting: Specifying Globalization and Resistance." In *Globalization and Resistance: Transnational Dimensions of Social Movements*, edited by Jackie Smith and Hank Johnston, 229–249. Lanham, MD: Rowman & Littlefield Publishers.

Tarrow, Sidney. 2004. "The Dualities of Transnational Contention: Two Activist Solitudes or a New World Altogether?" *Mobilization* 10:53–72.

Tarrow, Sidney. 2005. *The New Transnational Activism*. Cambridge: Cambridge University Press.

Tarrow, Sidney. 2010. "The Strategy of Paired Comparison: Toward a Theory of Practice." *Comparative Political Studies* 43:230–259.

Tarrow, Sidney, and Doug McAdam. 2005. "Scale Shift in Transnational Contention." In *Transnational Protest and Global Activism*, edited by Donatella Della Porta and Sidney Tarrow, 121–150. Lanham, MD: Rowman & Littlefield.

Terhorst, Philipp. 2008. "'Reclaiming Public Water': Changing Sector Policy through Globalization from Below." *Progress in Development Studies* 8 (1): 103–114.

Thomas, David M. 2008. "Past Futures: The Development and Evolution of American and Canadian Federalism." In *Canada and the United States: Differences that Count*, edited by David M. Thomas and Barbara Boyle Torrey, 295–316. Peterborough, ON: Broadview Press.

Tickell, Adam, and Jamie Peck. 2003. "Making Global Rules: Globalization or Neoliberalization." In *Remaking the Global Economy*, edited by Jamie Peck and Henry Wai-chung Yeung, 163–181. London and Thousand Oaks, CA: Sage.

Tilly, Charles. 2004. *Social Movements 1768–2004*. Boulder, CO: Paradigm Publishers.

Tilly, Charles, and Sidney Tarrow. 2007. *Contentious Politics*. Boulder, CO: Paradigm.

Tindall, David B. 2004. "Social Movement Participation over Time: An Ego-Network Approach to Micro-Mobilization." *Sociological Focus* 37 (2): 163–184.

United Nations. 1992. *Dublin Statement on Water and Sustainable Development.* International Conference on Water and the Environment (ICWE): Dublin, Ireland.

United Nations/World Water Assessment Programme. 2003. *1st UN World Water Development Report: Water for People, Water for Life.* Paris: UNESCO and Berghahn Books.

Urry, John. 2000. *Sociology beyond Societies: Mobilities for the Twenty-First Century.* New York: Routledge.

Urry, John. 2003. *Global Complexity.* London: Polity Press.

Vancouver Sun. 2001. "Privatization Can Help Keep Drinking Water Clean: A Huge Investment Is Required to Ensure our Safety." April 26.

Van Dyke, Nella. 2003. "Crossing Movement Boundaries: Factors that Facilitate Coalition Protest by American College Students, 1930–1990." *Social Problems* 50 (2): 226–250.

Voss, Kim, and Rachel Sherman. 2000. "Breaking the Iron Law of Oligarchy: Tactical Innovation and the Revitalization of the American Labor Movement." *American Journal of Sociology* 106:303–349.

Walker, Edward T., Andrew W. Martin, and John D. McCarthy. 2008. "Confronting the State, the Corporation, and the Academy: The Influence of Institutional Targets on Social Movement Repertoires." *American Journal of Sociology* 114 (1): 35–76.

Waters, M. Dane. 2009. "What Is Initiative and Referendum?" I & R Factsheet, Washington, DC, Initiative and Referendum Institute.

Weber, Max. 1905. *The Protestant Ethic and the Spirit of Capitalism.* Translated by Talcott Parsons. Introduction R. H. Tawney. Reprint, Mineola, NY: Dover Publications, 2003.

Webster, Edward, Rob Lambert, and Andries Bezuindhout. 2008. *Grounding Globalization: Labour in the Age of Insecurity.* Oxford: Blackwell Publishing.

Whiteley, John M., Helen Ingram, and Richard Warren Perry, eds. 2008. *Water Place and Equity.* Cambridge, MA: MIT Press.

Wilder, Margaret. 2008. "Equity and Water in Mexico's Changing Institutional Landscapes." In *Water and Equity: Apportioning Water Among Places and Values,* ed. Richard Perry, Helen Ingram, and John Whiteley, 95–116. Cambridge, MA: MIT Press.

Wilson, Jeremy. 1998. *Talk and Log: Wilderness Politics in British Columbia.* Vancouver, BC: UBC Press.

Zipperer, Ben, and John Schmitt. 2008. "Union Rates Increase in 2007." *Union Membership Bytes,* Center for Economic and Policy Research, January.

Zuberi, Dan. 2006. *Differences that Matter: Social Policy and the Working Poor in the United States and Canada.* Ithaca, NY: Cornell University Press.

Index

Urban and Industrial Environments

Series editor: Robert Gottlieb, Henry R. Luce Professor of Urban and Environmental Policy, Occidental College

Eran Ben-Joseph, *The Code of the City: Standards and the Hidden Language of Place Making*

Nancy J. Myers and Carolyn Raffensperger, eds., *Precautionary Tools for Reshaping Environmental Policy*

Kelly Sims Gallagher, *China Shifts Gears: Automakers, Oil, Pollution, and Development*

Kerry H. Whiteside, *Precautionary Politics: Principle and Practice in Confronting Environmental Risk*

Ronald Sandler and Phaedra C. Pezzullo, eds., *Environmental Justice and Environmentalism: The Social Justice Challenge to the Environmental Movement*

Julie Sze, *Noxious New York: The Racial Politics of Urban Health and Environmental Justice*

Robert D. Bullard, ed., *Growing Smarter: Achieving Livable Communities, Environmental Justice, and Regional Equity*

Ann Rappaport and Sarah Hammond Creighton, *Degrees That Matter: Climate Change and the University*

Michael Egan, *Barry Commoner and the Science of Survival: The Remaking of American Environmentalism*

David J. Hess, *Alternative Pathways in Science and Industry: Activism, Innovation, and the Environment in an Era of Globalization*

Peter F. Cannavò, *The Working Landscape: Founding, Preservation, and the Politics of Place*

Paul Stanton Kibel, ed., *Rivertown: Rethinking Urban Rivers*

Kevin P. Gallagher and Lyuba Zarsky, *The Enclave Economy: Foreign Investment and Sustainable Development in Mexico's Silicon Valley*

David N. Pellow, *Resisting Global Toxics: Transnational Movements for Environmental Justice*

Robert Gottlieb, *Reinventing Los Angeles: Nature and Community in the Global City*

David V. Carruthers, ed., *Environmental Justice in Latin America: Problems, Promise, and Practice*

Tom Angotti, *New York for Sale: Community Planning Confronts Global Real Estate*

Paloma Pavel, ed., *Breakthrough Communities: Sustainability and Justice in the Next American Metropolis*

Anastasia Loukaitou-Sideris and Renia Ehrenfeucht, *Sidewalks: Conflict and Negotiation over Public Space*

David J. Hess, *Localist Movements in a Global Economy: Sustainability, Justice, and Urban Development in the United States*

Julian Agyeman and Yelena Ogneva-Himmelberger, eds., *Environmental Justice and Sustainability in the Former Soviet Union*

Jason Corburn, *Toward the Healthy City: People, Places, and the Politics of Urban Planning*

JoAnn Carmin and Julian Agyeman, eds., *Environmental Inequalities Beyond Borders: Local Perspectives on Global Injustices*

Louise Mozingo, *Pastoral Capitalism: A History of Suburban Corporate Landscapes*

Gwen Ottinger and Benjamin Cohen, eds., *Technoscience and Environmental Justice: Expert Cultures in a Grassroots Movement*

Samantha MacBride, *Recycling Reconsidered: The Present Failure and Future Promise of Environmental Action in the United States*

Andrew Karvonen, *Politics of Urban Runoff: Nature, Technology, and the Sustainable City*

Daniel Schneider, *Hybrid Nature: Sewage Treatment and the Contradictions of the Industrial Ecosystem*

Catherine Tumber, *Small, Gritty, and Green: The Promise of America's Smaller Industrial Cities in a Low-Carbon World*

Sam Bass Warner and Andrew H. Whittemore, *American Urban Form: A Representative History*

John Pucher and Ralph Buehler, eds., *City Cycling*

Stephanie Foote and Elizabeth Mazzolini, eds., *Histories of the Dustheap: Waste, Material Cultures, Social Justice*

David J. Hess, *Good Green Jobs in a Global Economy: Making and Keeping New Industries in the United States*

Joseph F. C. DiMento and Clifford Ellis, *Changing Lanes: Visions and Histories of Urban Freeways*

Joanna Robinson, *Contested Water: The Struggle Against Water Privatization in the United States and Canada*